YOU ARE YOUNGER THAN YOUR AGE

— LAWRENCE LA ROSE —

BALBOA
PRESS

A DIVISION OF HAY HOUSE

Balboa Press books may be ordered through booksellers or by contacting:

Balboa Press
A Division of Hay House
1663 Liberty Drive
Bloomington, IN 47403
www.balboapress.com
1-(877) 407-4847

Because of the dynamic nature of the Internet, any web addresses or links contained in this book may have changed since publication and may no longer be valid. The views expressed in this work are solely those of the author and do not necessarily reflect the views of the publisher, and the publisher hereby disclaims any responsibility for them.

The author of this book does not dispense medical advice or prescribe the use of any technique as a form of treatment for physical, emotional, or medical problems without the advice of a physician, either directly or indirectly. The intent of the author is only to offer information of a general nature to help you in your quest for emotional and spiritual well-being. In the event you use any of the information in this book for yourself, which is your constitutional right, the author and the publisher assume no responsibility for your actions.

Any people depicted in stock imagery provided by Thinkstock are models, and such images are being used for illustrative purposes only.

Certain stock imagery © Thinkstock.

ISBN: 978-1-4525-8038-8 (sc)
ISBN:978-1-4525-8039-5 (e)

Printed in the United States of America.

Balboa Press rev. date: 2/6/2014

PREFACE

Affirmatively, information in this book will update the readers about their true age. The current calendar does not give a person their correct age, but increases it from year - to - year. The calendar is not giving an accurate time line in the creation of the worlds, and the many discoveries that were made by archeologists and scientists. Psychologically, the system is making one believes that their years are rapidly accelerating, and that they are not by any means achieving their maximum time effectiveness or their set goals. Even so, when women would have achieved the age of forty years old on the Gregorian calendar, they tend to believe that they are over the hill. In some instances, they may be discouraged by their loved ones in not taking the risk of conceiving a child. Things are not always what it is appearing to be. Many people looks younger than their age, likewise they feel that way in their body and mind. Below are a few examples in the various categories of age groups, showing that you are younger than your age according to the Prophetic calendar.

Gregorian calendar	Prophetic calendar
Present Age	Less young by:
23 years	5 years
30 years	7 years
40 years	10 years
50 years	12 years
60 years	15 years
70 years	17 years
100 years	24 years

There is a master key in unlocking the hidden truth about your actual age. It will be done by crafting a new Global calendar that has the correct amount of days in the year. A concentrated effort by politicians is urgently needed at this point and time in initiating a tangible solution, by devoting appropriate attention in accomplishing that task of reforming the Gregorian calendar. With forty days in each month, more recognized holidays can be added in satisfying the needs of the labor force and the schools. No other calendar has come close or parallel to the Prophetic calendar. More details are given on everyone's age in the second chapter.

Profoundly, whatever is disclosed in this book can be unpalatable, or it cannot be comprehensible to the average reader, because interest will not be shown by many in knowing the truth about God's creation, and others having an open mind- even the prominent scientific community.

This is not a science textbook or it is about any direct religion, but it is about spirituality. It is also an illustration of what mainstream scientists in their analytical thinking are attempting to understand how the sun, earth and the moon are functioning in the universe.

Scientists today should have fully diagnosed in the past century, that the main problem of the weather changing is embedded within our present calendar dating system, a calendar that is slowly drifting from nature, reality and humanity. The Gregorian calendar is giving an inaccurate time line in every aspect of life as well as other things in the universe. There will be a turning point in the history of the human race when a new calendar is adapted into the world. The Prophetic calendar is the missing link, and it shall connect the dots to show how the earth, moon, sun and the other world are working in conjunction. No other calendar has come close, or parallel to the Prophetic calendar of 480 days, since the beginning of creation.

Man and mankind are not responsible in any way in, contributing to climate change as written in many books by the various authors. The Climatologists, meteorologists, zoologist, marine biologists and botanist worldwide should have recognized ages ago, that the animals, plants, birds of the sky, fishes in the sea and ants are not in any way being affected by the weather. Therefore, they are continuing to function in their normal way of life. The animals, birds and other species are very comfortable within their immediate environment- and they are far beyond extinction. Mainstream scientists should be focusing on analyzing essential data - if it is available from the past century to accurately prognosticate the weather. To date, a representative from the office of the Vatican 's has yet to come forth in announcing to the world that the Gregorian calendar is going adrift, a problem that also occurred with the Julian calendar in 45BC or 46 BC.

An abstract of this book is to make man aware, that by using the lunar numbers of three hundred and sixty in order to calculate time line, calendars days and for navigational purposes are the wrong approach. Man's logical ways of thinking are not in synchronization with nature and God's ways, because Planet earth was perfectly created by Him.

This book is paramount to those who are spiritually motivated and is aware of their Creator's existence. Also, they may have a desire for a spiritual guidance by the Holy Ghost. It is a learning opportunity for disobedient youths to come out of the arena and be a vanguard to those unspiritual individuals- either them being young or old. The world is a circular stage set with over eight billion peoples or actors at the various levels. Some are chosen by God, while the majority is nominated by the Satan - the iconic character of many, in doing evil things to others.

The current magnetic needle on a 360 degrees compass has a three hundred and sixty degrees chart. This compass chart is not regarded to be a complete circle. The true compass chart is consisting of 480 degrees circle; and its compass needle is also pointing towards 360 degrees magnetic north.

In biblical times in the Old Testament, God gave warning to man for four generations [120 years] prior to the flood of Noah in the year 1656. Genesis 6:3. In our current generation, if man and mankind do not change their evil ways, governments, nations and systems will be totally collapsing. There is nothing hidden that shall not be revealed, neither covered, that will not be made known. Luke 12:2. God shall bring forth a seed out of Jacob and out of Judah - an inheritor of his mountains, and the servants of God shall dwell there. Isaiah 65:9.1

Humankind has somewhat fundamentally reinvented the wheel, that was working perfectly well with God's nature, and since at the beginning of creation. Now, mankind has added two more seasons in the calendar year- one season to the summer and another in the winter. The spring season added is approximately three months prior to the summer; and the autumn season is approximately three months prior the winter months. Psalm 104:5 states, that God had laid the foundation of the earth; and that it should not be removed evermore. In addition, the minerals beneath planet earth namely coal, oil, salt and water shall not be diminished, until the return of Yahweh and His purging of planet earth with fire.

The non- existing weather changes and other chaotic problems could be rectified, based on the mechanics and recommendations in this book, *YOU ARE YOUNGER THAN YOUR AGE.*

The Almighty God has given man and mankind a free will to employ a logical and rational way of thinking, in developing constructive ideas to the highest degree of originality and creativity. The Prophetic calendar has no complexities – it shall give rise to a new generation in a modern and a technological world by changing the face of history. Also, it will shape the future for humanity by becoming the life line for the farmers, in them increasing food production in feeding an ever growing population across the globe. Job 32:8; 38:36 states that a spirit in all men and the inspiration of the Almighty that gives man understanding.

Who set the stage of deceitfulness in the world? It was Satan that set a perfect plan to lead astray the world, by changing God's calendar from 480 days to 360 days. Satan chooses to do so after the death of his first born hybrid son Cain. Cain had died at the age of 360 years -270 Prophetic years. Some of Satan's reasons were to create a blindfold over the eyes of the future generations in the world, in order to cause confusion, hate and to disrupt God's time line in serving the creator at the various celebrations. This had started with the woman Eve in the Garden of Eden. In Genesis 3: 4-6, the Satan said to the woman, that she will not die if she had a warm friendship with him, and that she will be as gods knowing good and evil.

The latitude and longitude lines on the 480 degrees compass have an equal 240 degrees plus and minus 240 degrees in the north, south, east and the west with Jerusalem the center of the African continent is at the center of the earth. The scientific community would not lose any credibility, but may gain insights in developing innovative approaches towards new solutions and creativities, beyond normal expectations if they do not stonewall or suppress the truth - that is, if the least by looking at the possibilities in this book, *"YOU ARE YOUNGER THAN YOUR AGE."*

Gridding the world map with a 360 degrees compass was a very impractical application in determining the size and shape of the earth. The new invented 480 degrees compass is the appropriate instrument in accurately mapping the continents, countries, oceans and other locations on earth, including the Antarctica. Psalm 25:14 states that the secret of the Lord is with them that fear Him, and He will show them His covenant, as we are governed by time which is a gift from Him.

# $\mathcal{I}$NTRODUCTION

One of the most important things to be acknowledged is that what so ever is written in this book will be plausible, and it is solely based on biblical scripture, along with spiritual inspiration from the Almighty God. God formed the light and He created darkness. He makes peace and also creates evil. Isaiah 45:7; Amos 3

The Gregorian calendar system is actually increasing ones age from the time of birth to the day that person exhales their last breath. In a matter of fact, no calendar in use today gives a bona fide time line of creation, with a person's accurate age in anything that exists in the universe and on planet earth. This is including the age of all human beings, animals, plants, insects, rocks, bones or artifacts that were found by archeologists

In accordance with the chronicle of nations and throughout their recorded history, many have invented their own calendar. Some countries have been attempting to perfect their calendars, in order to observe their sacred festivals in forecasting events and traditional holidays on their yearly almanac. Nations have attempted to identify the core problems as to why the calendar system is going adrift after a period of years. Those nations so far have not analyzed or came up with new strategies as how to modify their calendar.

Changing the calendar's main purpose was to have it working in conjunction with the weather conditions, agricultural programs, religious observances, scientific events and other prescheduled activities. To this day no calendar has its perfection, or any human furnishing one, that will synchronize with God's appointed festivals and seasonal events.

In the past centuries calendars have been modified in order to satisfy the needs of governments. The calendars were used for religious purposes, and as an agriculture almanac, but most calendars of the various nations are convoluted and not unified.

The country of India is in the year 2069. All countries that have adapted the Gregorian calendar are presently in the year 2014. Iran is in the year 1392, and China's year of the snake is 4710. The Jewish year is 5774 and the Muslim year is 1434. The Prophetic calendar is in the year 5509, because it is working in conjunction with the Cosmos, the other worlds, and not independently. It accurately clarifies the pattern of festivities as cited on its calendar. It carries 480 days in a calendar year and 40 days in each of its twelve months. The calendar is accurate within its two seasons and festivals - along with its annual events. No added months, days or year are required in order to maintain its accuracy and consistency. A typical example in the differences in the time line between a Prophetic calendar and the man- made Gregorian calendar is given below.

All calendars that nations are using today are in disarray. The calendars are not working in conjunction with the Cosmos, the weather and the seasons on planet earth. They are constantly going adrift over a period of years, and there are no immediate solutions in place in order to rectify the problems. No proven methods or proper techniques over the past few centuries have so far been introduced by any commission or a panel in achieving or inventing a calendar, that can be consistent with the dates on an annually basis. Only the Prophetic calendar will be able to compute ones age accurately. This including the day you were born, and the hour that will not be changing; but it shall remain the same throughout your life time until death.

The Gregorian calendar computes an estimated time as a nine- month period in child bearing until the day of delivery. In some cases the doctor would notify the women, that the baby may be delivered two weeks before or two weeks after nine months. The doctor's time of estimation is not incorrect, because he is following the dates on the Gregorian calendar. The number of months or days on the Prophetic calendar that a woman bares a child prior giving birth is seven months or 280 days. This is in according to the Prophetic calendar system that carries 480 days in a year and 40 days in each of its twelve months. The calendar will move the world in a positive direction.

An important observant is that the Prophetic calendar requires no adjustments on an annual basis or centuries ahead, in order to function effectively and efficiently to the planners. They have no essential error pertaining to the biblical prophesies - either in the past or future events. The Prophetic calendar has the potential in becoming extremely accurate, as it was a recorded history in the event of Noah's flood. (See 400 years of oppression, and Jubilee Calendar)

Fundamentally, the dates on the Gregorian calendar does not affect the plants, animals, insects on planet earth, because they have an instinct knowing the changes in the seasons; and they are not being dependent on any calendar

The Messiah had ascended into the seventh heaven in the Lord's year in 3979. The most important point is that He shall be returning to reign as the King on earth. But, His primary task is to apprehend and imprison the devil, by taking him down to the bottom of the pit for a time of one thousand Prophetic years - one thousand three hundred and fifteen years on the Gregorian calendar. In His coming, one of God's seven spirits shall be armed with a key and an infinite chain. He shall be the only one who can lock and unlock the chain off the Satan at the appointed time. This event shall take place after Satan's thousand years have been completed in the pit. Revelation 20:1-3. Christ time line for His returning to earth in the Prophetic year is 5992. The time on the Gregorian calendar shall be in the year 2633 - that's if the calendar has not yet been changed to the Prophetic calendar system. In the far future the time period from the year that Christ had ascended into heaven, to the time of His returning to earth is 5992 – 3979 = 2013 years and one month on the Prophetic calendar. The time lines for the returning of Christ to be among His people are as follow:

PROPHETIC YEAR	GREGORIAN YEAR
Ascended in year 3979	Ascended in 1AD? (Anno- Domini)
Returning in 5992	Returning in 2633
Time-6992 - 5509 = 483 years.	Time, 2633- 2013 = 620 years

The time of Messiah returning shall be viewed by all at the beginning of Passover on the 14th day in the first month at twilight time - evening on the 13th day according to the Prophetic year.

The releasing of Satan from the pit will only be for a short period of time in the year 6992- one thousand years from the time he was imprisoned. The time period on the Gregorian calendar will be in the year 3633. There will be jaw dropping and awe in a later chapter in this book that will disclose the mysterious beast. The number 666 depicts that person who will be the devil's lieutenant. The beast will be severely wounded in battle, but he will survive to fight the final battle between good and evil at the end of times.

The above time line to the returning of the Messiah is philosophically and fundamentally accurate according to the dates on the Jubilee calendar. Revelation 20: 1-3; 1:18 states that one of God's seven spirits shall come down from heaven having the key of the bottomless pit and a great chain in his hand. He will be getting hold of the Satan, bounding him for one thousand years on the prophetic calendar; but 1,315 years on the Gregorian calendar if it is still existing. The Satan shall be set free for a short period of time, in order to establish his army to be engaging in the final battle between good and evil.

Clashes with the various religions and their festive holidays can be avoided. Nations must joined together to work harmoniously in adapting the Prophetic calendar as the unified calendar of the world, with avoiding costly or any unnecessary adjustments. The advantages of a change may assist farmers in scheduling, by having a sound dependable program without fearing weather conditions that may eventually ruin their crops. Realistically, farmers will make a substantial contribution to their nation by producing more greenery abundantly beyond normal expectations. That is, if the Prophetic calendar is adapted.

The weather is appearing to be a hindrance, and also incomprehensible to the human race, that is becoming frustrated and dissatisfied with the fluctuating weather conditions, instead of blaming it on the Gregorian calendar. Countries that are encouraged to adapt the Prophetic calendar may well foster a curiosity as to the mechanics, and its day to day functions. Most importantly, it may be useful in various developmental areas, and it will be considering innovative possibilities by nations producing their own greeneries and becoming self- sufficient; and not solely relying on global food imports.

The world's population is accelerating at a very fast rate, and nations should have a contingency plan in place to feed its people. Some examples of adjustment or changes in order to transform planning into a simple reality are as follows:

- Navigational compass [current 360 degrees]
- Radio carbon dating system
- Birth and death certificates
- Delivery date of a baby
- School curriculum extension
- Annual salary adjustments
- Annual scheduled activities
- Changes in the weather forecasting system.

Apparently, many other changes will have to be made with the following books because of their time line. These books will be: history; science; archeology; space travel; numerology; climatology; meteorology and many in other field of education – and those changes will have major interruptions.

Many debating within the educational system shall take place, because of changes in ninety percent of the world's present systems. It will be similar in changing a nation's economic, agriculture, scientific, including space travelling systems- and in many other areas in nation building. But to a nation's benefits and advantages, it shall be creating new jobs, new technology, or in some cases it may be having a domino effect on all systems- conversions. Definitely, it will cost an enormous amount of money in doing those conversions in many areas of science, space travelling and archeology. It will be affecting every aspect of human life on earth. However, transformation shall be promoting many creative solutions and devising means of accomplishing tactical ideas, new strategies; or excelling and enhancing ones performance in the educational system.

Some legal changes would be: voting; Jury Duty age limits; entering a bar; purchasing tobacco; enlisting in the military or paramilitary forces; driving and flying licenses; marriage license; Real Estate; purchasing of sea, air and land vehicles; graduating and teaching ages; gun and firearm licenses; participating in world sport events; practicing law and entering politics.

The purpose of the reform is to enhance our social, economic, scientific, cultural, educational, civil and international needs. It is a jump - starts in developing and creating new jobs in the field of computers, in terms of employing new hires working on the conversions of the above changes; and towards a nation's continuing improvements in educational achievements throughout the world. In particular, it shall be a game – change in the history of the universe, including every archeological discoveries on planet earth.

Humanity has evolved unknown to itself, under the increasing stress of a non-functioning and an artificial calendar system. It is true that man and mankind are operating under their own instinct and their logical ways of thinking, but not under the heavenly ways; and the ways of the creator of all things.

Rome once had the status of super power of the world with Emperor Julius Caesar introducing the Julian calendar in 46 or 45 BC. This calendar became inaccurate over a period of time, but it was replaced with the Gregorian calendar which was adapted by most nations in the year 1582; and it subsequently followed with other nations accepting the calendar. If accepted by nations in all continents of the world, and in the various languages, the Prophetic calendar may very well become the Global calendar. It shall be an instrument free of complexities with an avoidance of utilizing the sun or moon in calculating the months or the year for everyday use; and for a nation's administrative or agricultural purposes.

God did not create the sun and moon for the intent of man to formulate a calendar. He created them for all peoples and living organism on earth to benefit from the sun rays and the moon light at night. He appointed the moon for seasons; and the sun knows it rising and going down Psalm 104. All of God's festivals except the rest days were celebrated at the wrong time in biblical times due to a man- made calendar. Presently, God does not acknowledge the festival days that are celebrated. Isaiah 1:14-15

One of the main emphases in this book is based on the Prophetic calendar and its time line, which will give an accurate period of time to all things in existence. It is an innovative planning tool, and well synchronized with the vast universe. It is the key to understanding as to how the universe is functioning; and that man will be recognizing the rhythm of the constant recreating of the heavenly hosts. It will be capable of redefining the worldwide web system, by deciphering the various unbroken codes.

The Prophetic calendar's ultimate goal is to be calculating the true age of every human being, animal, plant rock, and everything else in the universe. It shall be the tool for resolving the mathematical problems - as to who is the beast that the bible is referring to in the latter days. Most importantly it will be predicting various changing conditions in the cosmos and on earth.

Pragmatically, people may not be accepting changes right away unless evidently convinced by the scientific community, biblical scholars, influential professionals and historians. In such case, changes would be facing many obstacles - even if they are leaning towards new perspectives, new technology, new trends and visions, but it would be a struggle in achieving those desiring results.

Changes will meet plenty of resistance in foreseeing some type of reaction, especially by the scientific community, or they may be having plenty of criticism, and "What if" scenarios as in the case of national elections, prior voting.

In adapting the Prophetic calendar, it will effectively control costs through economical utilization of materials, equipment, appropriate adjustments; and in other important areas of development on a long term basis. Innovative changes would establish a set goal for the improvement in the field of agriculture, science and a better educational system. It would be able to explore new opportunities and better technology. It will set clear and measurable objectives that can be counted on achieving lasting results, and maximum time effectiveness. The calendar can be used productively in identifying and eliminating wasted time. It would be an eye-opener - of course in innovative thinking for newly graduates.

Nations with a compelling international appeal to the European Union, the churches, the United Nations, the Muslim nations, the Jewish organizations, and other world bodies, may well convince the leaders of the free world to adapt the Prophetic calendar. Changes would have to be made within the school curriculums, both public and private schools. Visionary leadership in the educational system is expected to be properly executed when required, more so, in an event of a natural disaster as storms, hurricanes, floods or earthquakes. A natural disaster may force the closing of businesses, transportation system and schools. Solution to the problem may be resolved, if the Prophetic calendar is adapted by replacing the Gregorian calendar. The Prophetic calendar carries forty days in a month, and with the extra ten days it would become useful to the education system, that is, if students miss classes because of a natural disaster. The students would not have to make up time lost on their spring, winter and summer breaks, or losing out on vacationing with friend or family for a fault not their own. Job 37:9. Out of the south comes the whirlwind and cold out of the north.

CHAPTER ONE

LOSING TRACK OF TIME

The second temple of Yahweh was destroyed by the Romans in the year 4009. The children of Jacob (Israel) were scattered all over the face of the earth and far away from the treasured land, which was promised to their forefathers, Abraham, Isaac and Jacob. The Romans calendar at that time was the Julian calendar which was adopted in the 5th century. Then in the 16th century, the Gregorian calendar was introduced. The Gregorian calendar reformed of the Julian calendar was still in use at the time of Pope Gregory the thirteenth. The exiled people from the land of Israel did follow the dates on the Julian and the lunar calendars until this day. They lost track of God's appointed festivals, and lost track of precious time. Yahweh said that He hates the new moons and appointed feasts that were celebrated, Isaiah 1:14-15. The reason why is, because the festivals were celebrated on the wrong dates due to man made calendar.

With the emerging trends of all the natural disasters, the Gregorian calendar has become irrelevant to the changes taking place in the Cosmos, which is directly affecting humanity on earth. Man has also lost track of time on targeted dates on their basic annual events, including the weather forecasts related matters. The Prophetic calendar's only anecdote is by replacing the Gregorian calendar with the Prophetic calendar. The Gregorian calendar is ultimately responsible to a claim of changing conditions all over the world. It is the main reason why everything in the universe is dated inaccurately.

Some ancient calendars, such as the Babylonian or the Persian calendars use 360 days in their calendar year - it is known as the lunar calendar. The Gregorian calendar is conceived not to be a practical mathematical or geometrical instrument, in doing a good job to the satisfaction of nations in the world. There are no trends to follow when using the Gregorian calendar.

Every nation calendar differs regarding its beginning month of the year, and its ending month in the year. There is no uniformity with many of the calendars, and they have no standard of measure, that can be used to give an accurate time line. They all have flaws, and they are very frustrating to its users - primarily the farmers. The calendar programs a person's mind to accept what is proffered or what it is proclaiming. It is appearing that the entire world is comfortable with the system of irregularities.

Nations truly believe in the ancient crafters of the calendar system, even though those people as a nation do not exist on earth. The last inventors that invented the Gregorian calendar are derived from the Papal or the office of the Vatican. There can be in possession of all plans and documents, and knowing how the calendar was formulated. Bearing in mind that most people, especially the farmers have an adverse or unfavorable opinion as to how the system is functioning. Furthermore, the system is very impractical to many nations around the world. At the same time they are in denial of the situation and problems that the calendar is creating in every aspect of life including the age of everyone on planet earth.

The world would totally welcome, and they will very well appreciate an additional ten days to the thirty or thirty one day month, by making it forty days in a month calendar. The students would be able to accomplish more, that is, if the hours in the day are shortened to five hours. There is no reason to have classes on the student's winter, summer or spring breaks, because of no fault of their own, due to a natural disaster or snow storm. The extra time on a calendar of forty days can be properly utilized, rather than confiscating the student's time. On the flip side, class room discussions should be focusing on the creations of jobs, entrepreneurship, or science and mathematics, because the Almighty God creations are based on His specification of pure mathematical calculations.

Giving the true facts about the functioning of the Gregorian calendar's system, and where it was originated can remain a mystery to some people. The Gregorian calendar was validated by Pope Gregory XIII, and subsequently adapted in the middle part of the sixteenth century by other countries, and maybe it was unchallenged. *China was the last country to adopt the Gregorian calendar.* It is one of many man made calendars, that its system is mostly perplexing with many problems with inconsistency on a yearly basis. No explanation was given as to how the system functions, but is being alarmingly publicized in books and science magazines on situations or events of global warming and climate change.

The creator of the universe made the sun and moon for the purpose of lighting up the earth by day, and giving moonlight by night respectively. There were two instances that occurred in biblical times when God reversed the degrees of the sun on the sun-dial clock of King Ahab by ten degrees. **Isaiah 38:8.** . .

In retrospect as to how the Gregorian calendar was tabulated, the first six months adds up to a sub-total of one hundred and eighty one days. The other half of the calendar in twelve months is adding up to one hundred and eighty four days - an unequal balance as indicated on the chart. The following shows how the Gregorian calendar was put together.

One day, and night = 24hours
Seven days (one week) = 170hours
One month = 28 days
Twelve months is, 28 × 12 – 336 days
Thirteen months is, 28 × 13 = 364 days
The current Gregorian calendar in use has 365 days in one year and a leap year in adjustments of 366 days. The numbers of 365 days were derived at by adding an extra month of 29 days.

First half of the year	Days	Added days	Month Second half of the year	Days	Added days
January	31	3	July	31	3
February	28	(plus 1)	August	31	3
March	31	3	September	30	2
April	30	2	October	31	3
May	31	3	November	30	2
June	30	2	December	31	3
Sub Total	181	13	Sub Total	184	16

The Equation: Calendar days are 181 + 184 = 365 days. The second half of the year of twelve months has an unequal balance of days by three. A thirteenth month was omitted with no explanation given.

A true definition of a calendar should be as follows: A calendar is a system of organizing units formatted to compute and maintain year to year records consistently; to plan for systematic results; to be able to anticipate emerging problems, forecast realistic goals; to recognize or analyzes in tracking certain trends over a period of time; to make and project agricultural decisions; to worship the creator at his appointed times; to plan festivals, activities and religious events; to analyze the trend of history occurrences; to establish budget objectives and other fiscal matters; to initiate new ideas. It comprises of seven days called a week- 40 days in each month, twelve months consisting of four hundred and eighty days in a calendar year- it utilizes two seasons in the year depending on the geographic location of the land; it is a device used by a nation or society for agricultural, cultural, civil or religious matters.

# Chapter Two

YOUR ACTUAL AGE

Below are two examples of age Differences on the Prophetic Calendar compared to the Gregorian calendar:

EXAMPLE # 1

John is 50 years old on the Gregorian calendar, but he is actually 38 years old according to the Prophetic calendar.

Conversion:

Gregorian 50 year × 365 days = 18,250 days

Prophetic 18,250 ÷ 480 days = 38 years

EXAMPLE # 2

Mary is 100 years young on the Gregorian calendar, but actually her age on the Prophetic calendar: 100 × 365 = 36,500 days.

Prophetic years = 36,500 ÷ 480 = 76.

Lunar calendar: 100 ×360 = 36,400 days

Prophetic years = 36,400 ÷ 480 = 75.83333333333333

In calculating your age, the date of birth on the Gregorian calendar must be included, in order to have an accurate age on the Prophetic calendar.

The below biblical names are men who lived prior to the flood in the year 1656. They died at the following age, in accordance with the below listed calendars:

Name	Prophetic year's	Gregorian year's [Not biblical]	Lunar years [Works in conjunction]
Abel	100	131\ 5 months	*133.3333333333333
Cain	270	355	*360 (lunar)

Adam	930	1,223	1,240
Enoch	365	*480	486.66666666666667
Seth	912	1,199\3 months	1,216
Enos	905	1, 190\1 month	1,206.6666
Cainan	910	1,196\7 months	1,213.333333333333
Mahalaleel	895	1,176/9 months	1,193.333333333333
Jared	962	1,265	1,282.666666666667
Lamech	777	1,021/ 8 months	1,036
Methuselah	969	1,274/3 months	1,292
Rue	207	272/2 months	276
Peleg	209	274\ 8 months	278.6666666666667
Nahor	148	194\6 months	197.3333333333333
Noah	950	1,249\3 months	1,266.666666666667
Serug	200	263	266.6666666666667
Shem	500	657\5 months	*666.6666666666667

The following biblical names are the post flood years:

Name	Prophetic year's	Gregorian years	
Terah	205	269\5 months	273.3333333333333
Abraham	175	230\1 month	233.3333333333333
Sarah	127	167	169.3333333333333
Ishmael	137	180\1 month	182.666666666666
Isaac	180	236\7 months	240
Jacob	147	193\3 months	196
Joseph	110	144\6 months	146.6666666666667
Moses	120	157\8 months	160
Aaron	123	161\7 months	164
Meriam	93	122\3 months	124

Abel, Cainan, Mahalaleel, Nahor, Terah, Abraham and Sarah - all their ages at time of death had ended with the decimal numbers 33333……... It means that God did not place their spirits into another body. However, those characters that died having the decimal numbers with 66666…shall be reincarnated at the end of the age. They will play an important and crucial role, either good or evil on planet earth, prior the New World to come and after.

AUTHENTIC AGE ACCORDING TO THE PROPHETIC CALENDAR

The below information randomly highlight some interesting date of births when utilizing the Prophetic calendar:

Gregorian Age	Prophetic Age
2 years	1 year 5 months
3 years	2 years 2 months
4 years	3 years
5 years	3 years 8 months
6 years	4 years 5 months
7 years	5 years 3 months
8 years	6 years
9 years	6 years 8 months
10 years	7 years 6 months
11 years	8 years 3 months
12 years	9 years 1 month
13 years	9 years 8 months
14 years	10 years 6 months
15 years	11 years 4 months
16 years	12 years 1 month
17 years	12 years 9 months
18 years	13 years 6 months
19 years	14 years 4 months
20 years	15 years
21 years	15 years 9 months
22 years	16 years 7 months
23 years	17 years 4 months
24 years	18 years 2 months
25 years	19 years
26 years	19 years 7 months
27 years	20 years 5 months
28 years	21 years 2 months
29 years	22 years
30 years 6 months	23 years 2 months
31 years	23 years 5 months
32 years	24 years 3 months
33 years	25 years

34 years	25 years 8 months
35 years 6 months	27 years
36 years	27 years 3 months
37 years 6 months	28 years 5 months
38 years	28 years 8 months
39 years 6 months	30 years 1 month
40 years	30 years 4 months
41 years 6 months	31 years 3 months
42 years 6 months	32 years 3 months
43 years	32 years 6 months
44 years 6 months	33 years 9 months
45 years	34 years 2 months
46 years 6 months	35 years 4 months
47 years	35 year 7 months
48 years	36 years 5 months
49 years	37 years 2 months
50 years	38 years
51 years 6 months	39 years 2 months
52 years	39 years 5 months
53 years	40 years 3 months
54 years	41 years
55 years 6 months	41 years 8 months
56 years	42 years 5 months
57 years	43 years 3 months
58 years	44 years 1 month
59 years	44 years 8 months
60 years 6 months	46 years
61 years	46 years 3 months
62 years 6 months	47 years 6 months
63 years	47 years 9 months
64 years 6 months	49 years 1 month
65 years	49 years 4 months
66 years	50 years 1 month
67 years	50 years 9 months
68 years 6 months	52 years 1 month
69 years	52 years 4 months
70 years	53 years 2 months

71 years	53 years 9 months
72 years	54 years 7 months
73 years 6 months	55 years 9 months
77 years	58 years 5 months
78 years	59 years 3 months
79 years 6 months	60 years 5 months
80 years	60 years 8 months
81 years 6 months	62 years
82 years	62 years 3 months
83 years	63 years 1 month
84 years	63 years 8 months
85 years 6 months	65 years
86 years	65 years 3 months
87 years	66 year 1 month
88 years	66 years 9 months
89 years	67 years 6 months
90 years	68 years 4 months
91 years	69 years 1 month
92 years	69 years 9 months
93 years	70 years 7 months
94 years	71 years 4 months
95 years 6 months	72 years 6 months
96 years	73 years
97 years	73 years 7 months
98 years	74 years 5 months
99 years	75 years 2 months
100 years	76 years
105 years	80 years

In scriptures, God implied that man shall live three score and ten (70 years); but if man is strong and healthy he will see eighty years (80). Those ages are not calculated on the Gregorian calendar, but their numbers are kept on God's Prophetic calendar of 480 days in a year. Everyone of the human race has a number of years to live. Some may die at birth, and some may die young, while others may live to an old age. Human, humankind and everything living organism are born to die. Nine months on the Prophetic calendar is one year on the lunar calendar (480/360). Nine months and five days on the Prophetic calendar is one year on the Gregorian calendar (480/365).

CHAPTER THREE

THE INNER SECRET OF GOD CREATING

This chapter consists of a wide range of topics focusing on some important and essential matters, which are pertaining to most of God's creation within His seven worlds. The facts are supernatural experience, or intuitive basis of biblical knowledge of the scripture of things that constitute the Almighty God.

God is the creator of the heavens, the earth that was without any land- form, and with no human life existence, but only darkness was upon a flooded earth Genesis, 1: 1-2. In accordance to Genesis 1:2, the earth was flooded prior to its recreation. It was the first flood before the flood in the time of Noah, which was in the year 1656 to 1657 - and before the first physical Adam was created on earth.

When the Almighty God had the land submerged which was under water, it became one massive body of land known as an island, with the Garden of Eden in the easterly corner, and it was surrounded by one vast ocean or sea. However, God commanded the waters to abate, and some regressed to the underworld, while the remaining waters stayed abound. When God had ordered the land to submerge, the water separated itself from the land and simultaneously they both obeyed His commands. It is confirmed that the earth is a living planet, and namely *Mars, Venus, Mercury, Jupiter, Saturn,* etc. are barren planets which cannot sustain any sort of life on them without the spirit of God sanctioning them. In a matter of fact, mother earth played an important role in giving birth to man, animals, herbs and every living organism. God's singular and miraculous achievements are unquestionable by those who believes in creationism, but doubtful by the evolutionists.

No one should be in doubt that God is in control to the gateway of life, and that the Satan and his demons are responsible for the gateway to the graveyard. However, God uses the devil as a rod of correction, so as to get a message over to many or to an individual- or to a nation that needs redemption.

The Almighty is a God of seven spirits, and He is not seven particles. The spirits are in total control of the seven worlds along with what is beneath, and what is above the earth. The spirit can create things similar to matter, that it may cause a delusion to the human eyes, but matter or particles cannot create a spirit. Man is destined to have one spirit in his body. However, some people may be possessed with multiple spirits that will control their body and mind to do evil things, including suicide and murder.

In a few scripture verses, the God - Head had called upon the spirits saying, *"Let Us make man in our image after our likeness"* Genesis 1:26. Another example is in Genesis 11:7, when He said, *"Let Us go down and confound their language, that they may not understand one another speech."*

ULTIMATE GOAL

The purpose or aim of this chapter is not to debauch most scientific findings as to how the worlds were created or their expansionism, but in contributing by adding some imaginative, visionary along with logical thinking to the ordinary citizens, scholars or even to the scientific community. The biblical truth coupled with future explorations, may enable and encourage the scientific community to recognize the big picture of God's creating the heavens, the earth or beneath the earth. It is knowledgeable that eight billion living human beings and billion of animals breed in oxygen and exhale carbon dioxide. Mainstream science does not regard this as harmful to the ozone layer; or they are claiming that it is contributing to global warming. But in the case of factory, jet engine vehicle emissions, it is regarded by the scientific community as a major problem that is causing destruction to the ozone layer. CO2 or carbon dioxide is a necessary requirement for the survival of the plants and trees in the forests. They absorb carbon dioxide through their leaves, as it is written in the science text books. *Nothing is secret that shall not be made manifest, neither anything hid that shall not be known and come abroad. Luke 8:17.*

The rudiments in all the worlds, the earth, sun, moon and stars are working in conjunction with God's timing, measurements and nature. All things along with the hosts of the heavens are under His dominion and decorum, even though they are not of perfection, but they all worship their creator, except the *Satan*.

In Genesis chapter one, God had created them male and female, before he had created Adam and Eve. Later, God took a rib from the side of Adam flesh to form Adam's female companion who he had named Eve. God has divided the spirits of the good men and evil men, and it was later that the seed of Lucifer had emerged. God had inaugurated the spirits of the righteous or chosen ones at the time of His creation so that they will not sin - as it is written in the scripture.

In this world which is known as planet earth is populated with humans and humankind with flesh, that is designated to live for an average age of 92 years old (70 Prophetic years), mankind shall not accomplish that task if they do not desist from destroying themselves with nuclear weapons, illegal drugs, wars, sacrificing humans, and genetically modify food products. Scripture states that in the new world to come, a child may die at the age of one hundred years old, but the sinful ones will be a shadow, because they are very repugnant, and having no fear of the Almighty God. Ecclesiastes 8:12; Isaiah 66:24. Also, adults will live to a thousand years as it was intended to be in the beginning. However, mankind will continually to strive in attempting to change, reverse or slow down the aging process of the human race, so that they may live a longer life.

God is primarily a preeminent single root fundamentally formed with a mathematical equation originated from nothingness or zero. God is like a similitude to a *Roundtree* that has no beginning or no end, but an eternal and everlasting life. His true name originated from a formula in which He was the tree that was written in the scripture - alongside the tree of knowledge of good and evil, namely *Satan*. The *Satan* or *Lucifer* is known as the cunning serpent that allured Eve into committing adultery. Seven months later (280 days), she had conceived and brought forth Cain, her firstborn hybrid child. With one of God's compassionate attributes, man was given the opportunity to exercise his own free will - either to be righteous or to do evil, but man chooses the latter, which had resulted in a woman being beguiled by the serpent *Lucifer*. The word serpent which is mentioned in the scripture, is simply defined as the evil one Lucifer's personality and his behavior towards God - prior to him being booted out of heaven

God is the first, and He shall be the last. God was the word at the beginning of His creation, and He shall be that word at the end of all creations, before recreating the new heavens and the earth. The continuity of His heavenly council shall be relocated to planet earth on a temporary basis for one thousand years, until the new heavens are completed. God's members of His council are the twenty four elders, the seven spirits, the four beasts (lion, calf, man and eagle). Michael the arch angel is the fifth spirit of God. Revelation 16:10; 9:1, 2; 4:4 – 6.

The mega explosion at the beginning of the heavens was by His word He had uttered that created the heavens. When the first heaven will be passed, as stated in Revelation 21:1, The Almighty God shall destroy the worlds, but He will recreate new ones in an ongoing cycle. The recreation of planet earth is a period of seven thousand years, and the time line in the heavens is 1000 years to one day on earth. No dates can be affixed to the beginning of creation, because no one will be able to say when God came into existence – that is, creation goes in a cycle with no ending. He will recreate the new heavens with a mega- bang by an execution of His word. And nothing had existed before the spirit of God.

The heavens spreads out like a blanketed sequential shawl for all the peoples on earth to view the shining stars, by enjoying the beauty of God's creation that He is continually creating. Psalm 104:2 states that the heaven is like a curtain. Psalms 18:15; 135:7.

Some spirits that He had created are able to be manifested into flesh and retransform back into a spirit, because it was created to function in that manner; same as some of the hosts of angels that He had created in the heavens - and He who came to the earth as the Christ. Because of sin, planet earth is divided into a physical and spiritual dimensional world, while the other worlds physical beings vary in appearance from humans, the image of God and His hosts.

Conceptually, everything that is taking place on earth will not go unnoticed in heaven- the precepts are the same. God created the worlds in heavenly time cycle in a period of six thousand years - but the earth in six days.

The number seven has a significant meaning in creation of the heavens, the earth, including the sun. A cycle of Sabbatical or Jubilee years are period of time for change, when everything in the heavens and planet earth will bring about significant changes in the Cosmos. One earth year of four hundred and eighty days, is four hundred and eighty thousand days in the heavens or one year.

The heavens and the earth's time line of destructions and recreations are as follow:

The first cycle of destruction in the heavens had consisted of 140,000 Jubilee years × 50,000 = 7,000,000,000 years.

On earth, the time line of destruction is 7,000,000,000 heavenly years ÷ 1000 = 7,000 earth years. We are now in the first out of seven phases of destruction destined for the heavens and the earth. The recreation of the heavens and the earth are as follow:

Heavens is 7,000,000,000 billion years approximately, with 1,490,000 remaining years before the ending of the first cycle. Earth is 7,000 years approximately, with 1,490 remaining years before the ending of the first cycle. Because of sin at the beginning of the earth's creation, God had destroyed the earth by flood in the years 1656, but it had nothing to do with the cycle.

Mathematical Analysis of a cycle in the heavens is as follows:

The first 7,000,000,000 billion years in the heavens are one cycle, and a destruction of one heaven. Seven cycles are equal to one sabbatical phase of 49,000,000,000 billion years. The final destruction of the heavens and a new creation of other heavens will take place.

The last of Sabbatical Phases:

The first sabbatical phase is 343,000,000,000 (49 billion × 49 billion).

The second sabbatical phase is 49 × 49 × 2 = 686 billion years.

The third sabbatical phase is 49 × 49 × 3 = 1,029,000,000,000 years.

The fourth sabbatical phase is 49 × 49 × 4 = 1,372,000,000,000 years.

The fifth sabbatical phase is 49 × 49 × 5 = 1,715,000,000,000 years.

The sixth sabbatical phase is 49 × 49 × 6 = 2, 058,000,000,000 years.

The seventh sabbatical phase is 49 × 49 × 7 = 2,401,000,000,000 years.

A complete circle of the heavens to a Jubilee destination and changes are 49 billion × 49 billion × 200 = 480,200,000,000,000 years (4.802e+23). Planet earth will be 480,200,000,000 years old, because the heavens are one thousand years to one day ahead. Then the earth is moving along in synchronization with the heavens at 480 days in a year. Therefore, a chariot (UFO) from other worlds timing period cannot be measured by man. While the earth is rotating, the heavens are also in a state of continuing rotation, changing and moving in a spiral formation – similar to planet earth that is rotating at 960 mph in 24 hours. Planet earth is God's footstool, which is directly connected with a plum-line to the center and at the bottom of the earth

In the process of God creating the new heavens for the second time - His place of abode shall be on planet earth among His chosen people, for a period of one thousand years. *Revelation 20: 6 states, those that are raised from their graves will be in the first resurrection, and they shall reign with God for a period of one thousand years.*

In the second creation of the heavens and the earth – the moon will not have its cycle of twenty one years, but it shall be stationary in the solar system for man to celebrate every beginning of the month of forty days. Because of the elliptical gyration of the sun travelling at 3,500,000 miles an hour (immeasurable) around the seventh heaven, and planet earth at the end of the second age, it shall cease to function as a light to planet earth - and the sun will be voided. *In Revelation 20:21, states that the first heaven and the first earth were passed away, and there was no more.* The sun will be placed on an axial orbit - and not a revolving around the heaven and the earth. It shall be covered by thick dark clouds for a resting period of many years.

IN THE BEGINNING

ONGOING CYCLE

PLANET EARTH
SABBATICAL YEAR
49

7
COMPLETENESS

ROUND TREE
GARDEN OF EDEN

50
JUBILEE YEAR

0
THE SPIRIT OF GOD

ONGOING CYCLE

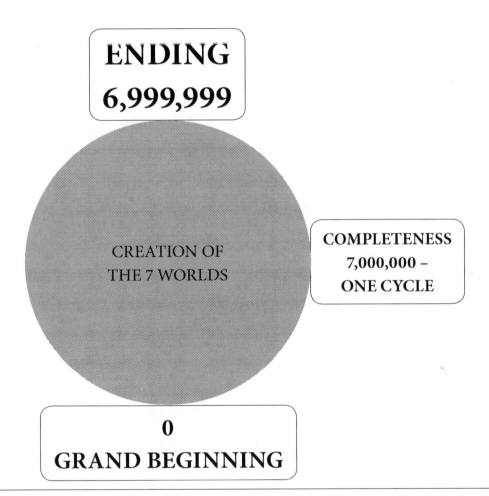

ENDING 6,999,999

CREATION OF THE 7 WORLDS

COMPLETENESS 7,000,000 – ONE CYCLE

0 GRAND BEGINNING

Genesis 1: 7, 8. God made the firmament and divided the waters (on earth) which were under the firmament (earth) …… and God called the firmament heaven, and it was the second day. There are six other worlds. Currently, God is dwelling in the seventh heaven, and the earth is His footstool. Nothing is at a standstill in the universes. Everything goes round and round - same as the new worlds to come, the new earth, a New Jerusalem and very good people, that shall be living in peace and harmony among the tribes of Israel and the gentiles. Earth's recreation cycle is every 7,000 years, and the heavens are 7,000,000 million years.

LAND PORTIONS OF THE TRIBES OF ISRAEL

Westside

↑

GAD
ZEBULUN
ISSACHAR
SIMEON
BENJAMIN
JOSEPH (double land portion)

Not drawn to scale

Sanctuary   **25,000 cubits**

10,000 CUBITS

LEVITES   *Northside →*

10,000 CUBITS

JUDAH
RUBEN
MANASSEH
EPHRAIM
NAPHTALI
ASHER
DAN

Not drawn to scale

Eastside

The city dwelling will be 25,000 cubits × 5,000 cubits

Going out of the tribal areas, east, west, north and south will be 4,500 cubits each way.

The size of the Temple and the size of the Levites portion of land will be 25,000 cubit × 10,000 cubits each

The entrance of the Levites is at the north gate.

The entrance of the Temple is at the north gate

The entrance of the Sanctuary is at the east gate

The entrance of the tribe of Gad is at the west gate

The entrances of the tribes of Zebulun, Issachar and Simeon are at the south gates.

The entrances of the tribes of Benjamin and Joseph are at the east gates.

The entrances of the tribes of Judah and Ruben are at the north gates.

Ephraim Naphtali and Asher entrances are at the west gates.

The entrance of the tribe of Dan and Manasseh are at east gates.

PROGRESSION

God's order of growth or changes in His universes and on the earth is carefully guided within a slow process of 7,000/7,000,000 years - similar to a continuous growth of a tree that produces branches, its bud, leaves, flowers and fruits.

Mankind is deliberately interrupting the natural weather cycle on the earth - maybe their reason is to take full control or to have total dominance of God's planet. However, the forces of the Almighty God will unknowingly to the human race - and in His own time shall bring about many changes to every valley, mountains and hills which shall be flattened like plains. Isaiah 40:4. In Isaiah 41: 18, it is concurred that He shall turn the valleys into rivers; and He will plant trees in the deserts. There shall be new valleys, new plains, new deserts, new lakes and oceans and with a New Jerusalem, standing tall high up on a mountain. All people will be able to journey to see and worship the Messiah in His new created Temple in Jerusalem. This Temple shall not be built by mankind or by hand. Any other temple that will be constructed will never be recognized as the house of God - but it would be intended for the house of *Ha Shem (the name)* and the *Satan.*

God is the truth, peace, with righteousness and is in control of everything that He created. God has favored His selected saints to be in His council in heaven. God expects His creation to utilize their intelligent minds to the fullest, solely for the benefit, also development of their habitat and the human race. He is allowing them to venture far beyond their immediate environment. All substantial developments or successes are expected to be conducted without any evil thoughts - and it must be pleasing in the sight of God. These are some of the validations of creation and His establishment in the heavens and upon the earth. He created the trees, human and humankind, same as the stars in the heaven - and subsequently they are eradicated in order to reproduce new ones by His sun.

Sincerely, the God of the universes has already been preparing a way for the new heavens and a new earth's recreation as spoken in Isaiah 65:17 and Revelation 21:1. Man and mankind inability to prevent or guard from the refashioning of the universes shall be futile in the latter days. Any occupation on any planet other than the earth shall never take place. Because, some of those barren planets may have been once inhabited by living creatures or beings millions or billions of years ago, but they were destroyed by the Almighty God, same that will be taking place in the heavens and the earth in the year 6999, 1,490 years from 5510. The Creator of the heavens and the earth can shorten the days if deemed it necessary

Psalms 8:3; Isaiah 40; 13-14. God is all about love, ethics and moral conduct. He gives man choices and reasoning in making their own judgment. He answers to no man or humankind. He provides food for every living organism on earth. In some instances

He allows the devil to have his own way with the human race that came from the seed of Adam. God is always aware of the devil's wickedness, but He is still in control of the Satan at all times. God forgives one's iniquities - only if they do sincerely ask Him for forgiveness. This is one of His many attributes that He possesses. In the beginning God had selected his chosen souls. Their faith has been sealed and their names are written in the book of life. At the end of time, God shall refine His chosen ones - the same as refining white gold, and they shall be transformed into angelic beings.

According to the experts in the field of astronomy, planet earth is revolving around the sun at 67,000 miles an hour. It is estimated that the earth covers 940 million kilometers or approximately 584 million miles around the sun. In calculating the time that the earth in completing one revolution around the sun, it should be as follow according to the number of miles given by the expert astronomers: $584,020,718 \div 67,000 = 8716.727$ hours. However, $8716.727 \div 24$ hours is equal to 363.1 days, with 1 day and 15 hours short of 365 days. The current compass has a circle of 360 degrees. A 360 degrees circle will be giving a maximum time of 8640 hours(360×24 hours); and a 365 days will be 365×24 hours = 8760 hours. A 480 degrees compass with a complete circle will be 480×24 hours = 11,520 hours.

At a distance of 584,020,718 miles that the earth is claimed to be going around the sun divided by 23,040 miles - the size of the earth will be approximately 25.3 billion times earth's circumference. So then, according to the scientists, earth will be travelling around the sun at a rate of speed of 69,440 mph. (25,347,220 miles divided by 365).

In mid- February 2013, an asteroid travelling at a rate at approximately 40,000 miles per hour - hitting some part of Russia and injuring many people. So then, if the earth is travelling at 67,000 miles per hour - there is no way that an asteroid travelling at a speed of 40,000 miles per hour would be able to make contact with planet earth. Therefore, the theory proclaimed by scientists, that planet earth is revolving around the sun is surely not the case.

A PLAUSIBLE VISION / GOD'S ATTRIBUTES

Fundamentally, God is the author and creator of everything. He is the composer and conductor in the rhythm of things in the worlds.

In the month of May 2004 in the wee hours in the morning, I had a clear vision that channeled me directly in the presence of the *Creator* who is residing in a huge mansion. Before entering His room, I was confronted by a female. She was standing approximately three feet away from me, and then I found myself in the kneeling position with the left hand outstretched. The woman placed a ring insignia on my left middle finger. Immediately I was at the standing position. She told me that my name is *Roundtree*, and that *Roundtree* is the richest ever. She said that wherever I go I will be recognized. Then I saw the woman standing at the back of a female approximately fifteen feet right of where I was standing and who I had recognized. She tore down a top portion of clothing at the back of the woman, and immediately, I noticed her in front facing the woman. With a low tone of voice, she told the woman, that her name was also *Roundtree*.

In my vision I witnessed that He is a spiritual being, that no one can estimate his true years of existence. God is very awesome, humble and benign in appearance. His hair is pure wool-like and comes along with a bang. He sits in a chair that has an arm-rest. The chair that he was sitting in had been levitated at approximately eighteen inches off the floor. It then turned clockwise around facing in my direction where I was standing - that is a few feet from the entrance to His room. His garment is lily-white with no blemishes. His smile was penetrating to my soul, and it was a satisfactory smile. His stature is petite, and He is about five feet in height, according to measurement of the size of chair that He was sitting in. He has beady- eyes and his eye brows appeared to be off white. His complexion is very fair – as a Caucasian, and He is clean- shaved.

God's female helper is also a Caucasian. She is approximately five feet ten inches in height and her hair drops down shoulder length. Her age appears to be about forty to fifty years old. Her demeanor and bearings are business- like and very officious, and somewhat authoritative. Her appearance in face, hair and other features befits a Caucasian woman. She had stopped short of entering a room that God alone had occupied.

The ceiling of the mansion is very high with the hallway measuring at approximately twenty feet wide and endlessly long. His room size is about twenty five feet by twenty five feet in width; and there was no door leading from the hallway to His domain. The entire space visible was very commodious.

THE SCIENTIFIC COMMUNITY PERPECTIVES

Science is considered to be a tedious task to the mainstream scientists. They have a wide spectrum of knowledge, due to the availability of new and advanced technology and ancient documents at their disposal. They somehow gave credit to themselves with having the abilities in achieving set goals in accomplishing their objectives. Prominent scientists may have more of a scientific approach as to how the universe was formed. It is rather than a religious philosophical approach as most ordinary people believes in a creator. Other scientists may have separate views on a subject matter or a discovery. Mainstream scientists may go to an extent in suppressing the truth on something that is presented to them by an amateur, a religious scientist or an outsider. Some scientists may eventually discard a new discovery if it is not in line with their theory or claim. Some scientists may be pragmatic, and they may not be believers in the Almighty God who created the seven worlds, but they will give Him a back-hand compliment. Those scientists do not have the spirit of God, but somewhat an ungodly spirit, that give them an inspiration to do things contrary to the word of God. A matter of fact, science claimed that man derived from the family of apes, but science fell short in denouncing that the ape was created by God. Humans are not just anatomy, or they have descended from the family of apes, but they were created by God from the dust of the earth. The spiritual images of the creator along with His hosts are not apes - but they are the image of the human race. There is a vast difference between man and ape. That distinction is that man has a soul, but the ape does not have a soul. There are no controversies, but it is only an axiom of creation. Can man make an ape speak the languages of the human race? These claims by the scientific community that man has derived from the ape family are oxymoron. Same as the previous generations that were told the earth was flat, but subsequently, that theory was recanted by the scientific community in concluding that the earth was round. *To date, mainstream scientists are now claiming that the earth is shaped like a sphere and bulging at the pole.*

Science has a different interpretation or perception as to the functioning of the earth - and the reasons for the natural disasters that are taking place in the world. The scientific community may describe, or may be giving an explanation based on their data of an instrument that measures earthquakes, and the magnitude of an earthquake. However, they cannot give tangible reasons as to why it is cyclically happening to certain nations; and at the same time natural disasters are not badly affecting others nations, even though their borders are connected. A typical example of the last hurricane *Sandy* in the month of October 2012 in the USA was not because of a change in climate, but it was destined to hit planet earth. A reason for drawing to this conclusion is, that I was shown in a vision, a very high tidal waves coming in from the ocean unto land, prior the disaster happening. This dream was repeated a second time in 2012, a few months before it took place.

The ordinary person should make an effort in getting out of the ring of fire, by having that self- perception in their own ability in understanding God's creation. They should do some researching on their own into the word of God.

Mainstream scientists cannot substantiate or confirm that the heavenly space is consisting of other worlds; but they may envisage a comprehensive theory in not being aware, what lies beyond the right hand of the throne of God who is residing in the seventh heaven.

Science explores new pathways, valuable insights in discovering new approaches. They have a solid, but also a strong power of observation in seeking solutions to a wide range of things. Scientists have much improved health programs and development in the field of medicine for the betterment of humanity on planet earth. Through their continuous studies and perseverance, the scientific community has gained conventional wisdom and knowledge by adding a broad, but a consistent application and dedication, that they are now displaying and recognized throughout the world. Scientists may be doing researching without a bias mind and with lots of certainties and uncertainties. The whole truth may not be revealed at all times of their new discoveries, but scientists may sometime delay their findings - or in keeping an open mind by not wanting any contradictions with their initial discoveries.

Science is about skepticism, speculations, false claims, retractions (as previously thought), theories; predictions and probabilities that must be verified by the scientific community. Science does not educate human being about love or moral conduct. They cannot differentiate evil from non – evil, but they are a few caricaturists that would like the world to believe in their publication or television programs - and about the creation of the universe. Can science prove or define the purpose of creation, or the expansion of the universes or worlds?

Science cannot answer to everything about the universe which comes under the umbrella of the Creator. They may identify over a period of time His magnificent works of creation, in attempting to figure a way as to how the universes are functioning. Ecclesiastes3:11.

In order to achieve any scientific results, an individual or a team of scientists will have to prove an experiment. In doing so, the project may have to be repeated by an individual or by team efforts. The experiment also has to be approved by the scientific community, so as to be successful in their endeavors for publication or funding

HUMAN COMBUSION

Has the science community ever claimed or discovered that the body of an ape had never gone through the process of combustion as the human body does? *The fact of the matter, it has been documented that some humans mysteriously had died from combustion, which caused their entire body to be burnt to a char.*

The human body carries a small amount of electricity commonly known as static. Theoretically, some people may have a build- up of excessive electricity in their body over a period of years. This may be caused by too much dirty electricity in the air, and maybe inherited from their environment or immediate surroundings. Also, the electrical installation that was done to the very old houses years ago were not fully protected from dirty electricity.

Modern houses are now being built and installed with electrical panels with well- grounded wires and circuit breakers. The old houses electrical system had carried antiquated glass fuses that were screwed unto a back wooden board affixed to the wall. When too much electricity is conducted to a single fuse there was no shut off system, in order to prevent an escape of electricity or in causing a fire- *that is known as dirty electricity.*

Coherently, it is reasonable to ask these illustrious and fundamental questions, that the mainstream scientists may be able to furnish the world with some truthful, but sincere answers:

By what way is the light parted, which scatters the east wind upon the earth? How can a tumultuous wind be blown if the sun is immobile?

From which direction at the cardinal point is the wind originating?

Has the wind ever come from the west to east, or south to north, other than a whirlwind? Is there any substantial or daily observation data that a stationary sun may be able to spurt out winds, so that a very powerful winds may reach planet earth?

Can the ingenuity of the scientific community produce any type of living organism in order to populate a planet with humans, animals, plants, birds, insects, fishes and having an ongoing parturition on that planet?

A logical and accurate assessment on the directions of a solar wind can be determined by the high rate of speed, which is created by the sun during its gyration around the one universe. The misinterpretation by the mainstream scientists in relevance to the functioning of the sun is not in line with the blue print specifications in the scripture.

The heavens, including its host along with the stars are full of beautiful music and pleasing to the Creator. They all worship Him at the appropriate hour. He knows each star by its name and the quantity of stars in the heaven. Everything in the universes and on earth has a connection with the seventh heaven where the Almighty God is located. Even the human race, humankinds, animals, birds, trees in the forest, fishes in the sea and insects have a connecting frequency that is somewhat imbedded into their system. Those frequencies are electromagnetic connections from planet earth and with the Cosmos. Some typical examples are the synchronization of the birds that are maneuvering in the air, and the synchronization of the movements of fishes in the sea.

At times, some people may display a type of static when touching a metal object or even by kissing someone on the lip- this maybe a negative energy that transcends from the other person. Everything on planet earth does contain a minute amount of electromagnetism which is dispersed throughout the worlds.

Job 37:12 states that the cloud is turned round about by God's counsels, that they may do whatsoever He commands them upon the face of the world in the earth. Psalm 148:8 fire, hail, snow, vapors, stormy wind fulfilling His word.

HUMAN AND HUMANKIND

The Creator of the heavens, other worlds or the earth has created the human race and humankind with inspirational ideas, innovative desires; and enthusiastic spirit to venture beyond their earthly based environment. Broad spectrums of the alien races arc from othcr spiritual worlds. They are spiritual transforming themselves temporarily on earth into humankind, and adapting themselves to the environment momentarily. The scripture described them as a curse of the earth, roaming to and fro throughout the worlds. They are being witnessed in many countries, and by many people around the world. They are described as UFO's or un-identified submerging crafts from another dimension below the earth's ocean, and oceans below the earth's crust.

Human beings have not so far achieved such advanced technology in terms of determining measurements, velocity, or their reasons for visitations. Those beings dematerializes themselves out of sight within a flash of light - the reason being so, is because the heavens above are four hundred and eighty thousand years ahead of planet earth.

Fundamentally, it is conceivable or inexplicable that beneath the earth may dwell strange or unfamiliar beings. Some of them are surfacing, and they capable of monitoring humans and humankind on earth - or man's nuclear activities, our new technology and other new developments on planet earth. Humans and mankind dwelling above the earth's crust maybe considered primitive to those beings beneath the earth and the other worlds. They are more technologically or intellectually advanced in space travel than humans. This assertion is based on the type of crafts or chariots that they pilot around the worlds. Often, they may submerge in the high seas on planet earth in an unknown sea or space- crafts, same as those crafts that are approaching from the heavens, which are described as UFO's. Philippians 2:10; Isaiah 44:23; Zechariah 5:1, 2 Ezekiel 1:13.

The extra- terrestrial beings which are dwelling beneath the earth may be able to communicate with humans through mental telepathy. *In Zechariah 5:4 states that the curse shall enter houses of the thieves; and to those that swear in the name of the Lord.*

The following are examples with the difference between human and humankind:

HUMAN	HUMANKIND
Has the spirit of God	*Created with the spirit of Lucifer*
Some are chosen	*fallen angels within some bodies*
Gentile nation	*Terrestrial beings from the underworld*

Bona- fide and Christ-like   ***Wolf- like (hybrid)***
                                          Evil spirits adapting human form
                                          Other beings in various forms

Humankind, including the ancient people that had previously existed on planet earth, might have emulated some scripture verses from the holy bible about God's creation and Him creating. Thereafter, they might have come up with their own interpretations as to how the earth was formed, or how humankind was created; and they also crafted a lunar calendar that was based on the appearances of the moon. Those claims had originated from their primary ancestor Lucifer, *the Satan*. It was then passed down to his generation, the Canaanites, and subsequently unto the Babylonians about three thousand years later. This took place after the death of Cain in the year 360. This time-line is based on the Canaanite's lunar calendar, but on the Prophetic, the accurate date of his death is 270 Prophetic years from the time of creation.

In Genesis 4:16-24 stated that Cain the first hybrid who was the son of Lucifer, and was banished from the presence of God. Cain then proceeded to dwell in the land of Nod - the land of Nod was located east of the Garden of Eden. He later then settled in the land of Nod permanently. He gave his people similar names to the Adam's generation. A couple of those names were Enoch and *Lamech*. Other similar names were given to his generation with the pronunciation being the same, but their spelling was different from Adam's generation – this is according to the English translation. Those strategies were intended to confuse the children of God with the generations to follow. They were lured, or deceived into believing that the names of *Lamech* and Enoch, followed by other deceptive names were genuine names of their families, that belonged to the Adam's generation. To this day, God's chosen people who were caught in the snare are unknowingly serving Lucifer, and they are naïve about it. It would be useless in attempting to change their philosophy. With their upbringing, tradition, and their teaching that they received from their elders, they will not deviate. That's why God is pleading to His people of Israel or Jacob, in saying that they must come out of her, because at the end of the age, He shall destroy every last one of them (*Edomites).* Their flesh shall be dissolving off their skeletons.

THE PARA- NORMAL

Allegorically, it is appearing that the monopoly of evil is speedily increasing in the super natural realm. They are possessing or they are depressing the human minds and bodies which are only affecting those who are making themselves vulnerable - like the drug Abusers and alcohol addicts. Those who are meditating towards the evil side other than meditating unto the almighty God would also be targeted by the demons in becoming their victim. They are unknowingly creating an opening in which demons may enter and take over their minds and bodies. These spiritual being takes control of a person, and in some cases causes them to become fearful, commit mass murders, suicide; or in displaying other negative acts in having no regards for human lives or empathy in their wrong doing.

Mankind known as humankind or hybrids has no sense of morality, but they would maim or kill anyone; and in the end they show no remorse in doing just that. Other evil spirits which are described as demons or ghosts are earthly bound. They love to dwell in dry or abandoned places, and also in the darkest of places. A demon or demons commandeer and depresses one's mind causing a person to behave in an abnormal manner. Some eccentric or psychotic behavior is as follows:

- Having a conversation with someone seen by only them, but not seen by the normal person passing by.
- Displaying a cold, unusual stare down in a menacing manner, with the pupil of their eyes becoming dilated and glossy or dark; or they may be very dogmatic in their comportment.
- Having a type of very low- key voice.
- A possessed person may uncharacteristically be driven to hatred against their loved ones. However falling asleep with a television on with it showing an evil program is not a very good idea, because evil can manifest themselves in that room and into ones subconscious mind. So then, between asleep and being awake the spirit may depress the body to an extent, that a person may have a temporary paralysis. They may even be unable to utter a single word; though their eyes are opened and they are conscious about their immediate surroundings.

The scientific community or medical journal characterizes erratic or strange behaviors as mental illness, bi-polar disorder, or the medical terminology as schizophrenic when a person is behaving in an abnormal manner, or when their personality has been altered

Demons have enormous power to levitate a human's body off the bed or even rocking the bed. They having power to toss objects around the room or rattling the kitchen utensils in the plate rack. They also have a ranking system as a high ranking lieutenant to low ranking subordinates. Demons have the tendency of attacking women while they are asleep. They at times leave the women with black and blue marks mainly in and around the region of the legs. If someone has a demon lying dormant in their body- for a fact they may not know that they are being possessed.

A tropical or South American bird known as a parrot may discern that evil spirit in that person in displaying an unusual behavior by flying around inside their cage, when that possessed person is approaching the cage. Even other animals like the dog or cat may be able to sense evil around.

It is possible that a spirit or an apparition may impersonate a loved one that has passed away. Also it's a possibility that those loved one who has passed, and who was very close to that individual may give a presage of an imminent danger or specificity of a situation, either in a vision or a dream. A person may describe such episode as saved by a guardian angel; or will say that they were visited by a loved one.

Some humankind practices a temporary leaving of the spirit from the body. Then the spirit would have the power to journey from point to point and later returning to that body (Astra- projection). Other negative influence affecting the spirit and mental state of mind are portrayed through the Media, music, video games, websites, arcades, movies, and comic books with evil characters or evil television shows.

Another experience of evil happened on a Friday night at the beginning of the Sabbath. Some invited guests were present in the house celebrating and taking part in the Kiddush and Sabbath meal. The adults were in the living room, while the two kids were playing the video games in the bedroom. When the guest went into the bedroom to tell the kids to cease playing the game and to join the adults at Kiddush, the person saw a Caucasian male sitting with the two kids. The guest came back out to the living room, inquiring if there was someone else in the house. The guest was told no, and the kids were immediately hurried out of the room.

Evil may manifest itself into a person thus causing them to have a split personality, a change in personality or many personalities. One of those personalities may affect them emotionally; or they may be displaying some kind of strange behavioral traits. *Here is a typical example of such behaviors: A person may create a scenario of a kind by making verbal complaints, that they are being stalked or threatened. They may enforce that allegation with written letters or notes left in some place visible to others. They may scrawl indecent words on the wall or elsewhere, so that it can be noticed. They at times will be using the other hand, so as to disguise their regular handwriting.*

The Satan is considered to be an angel of light. He has the power of creating delusional scenarios by projecting something that is not real or authentic. In Mark 13:22 states that false Christ and false prophets shall be raised showing signs and wonders, so as to captivate the attention of the mass.

A typical and living example of such a case took place in 2009 in the Bronx in the city of New York. Three young people had experienced evil in their home. They had installed a Jewish mezuzah on their apartment door frame on the right post. A young teen age boy had a 240 degrees turn around in his personality from being good, in becoming violent to his mother and very destructive in the home.

One day the mother and daughter had witnessed that the ceiling in the apartment had appeared to be cracking, the walls and floor were shaking, but actually they were not, and it was only a delusional spell handed down to them from the devil. Subsequently the mezuzah was taken off the door frame. It was then tossed away across the street and away from the entrance of the apartment. A few minutes later the doorbell began ringing non- stop. Another witness was in the apartment, and when they looked outside from the window they saw no one, but the bell continued to ring and eventually it stopped ringing. In 2 Thessalonians 2:9, 11 states that even him, who's coming is after the workings of the Satan with all power and signs and lying wonders. This will cause God to send them a strong delusion that they should believe in a lie.

Another case of witnessing an evil scenario was at my home. On returning to my house in the dark hour one evening, I saw a red light glaring on the right door post where a mezuzah was installed. As I got closer to the house, the light disappeared. When we entered the house which was in total darkness, there was a flash of light as if someone was taking a picture with a flash camera, but there was no one else in the house. We both were perplexed and astounded. In Isaiah 45:7 God said that He formed the light and that He created darkness; He makes peace and He creates evil.

A friend may encourage their peers to do acts of evil such as experimenting with an Ouija board that will be giving telepathic or spiritualistic messages. This curiosity can cause demons to enter that home or may find that individual and the appropriate body as a place to rest permanently. However evil can be rid of if one has a very strong belief in Christ as their savior or God as the Almighty of the worlds by maintaining, that spirituality without wavering, backsliding or being contiguous in such evil things. So then, one must solidify their alliance with the Creator of the universes.

I had experienced the presence of evil when I was in the process in writing my book. This all took place during the wee hours in the morning, when things were very quiet in the neighborhood.

At the entrance of my apartment, there was a mezuzah installed on the door. There were weird activities that were taking place in the apartment while I was awake, and at times when I get up in the mornings.

In the hours of darkness, there was a strong smell of fresh roses or the scent of perfume, even though I never had a female in my apartment. At times the bag in the garbage container would make sounds of things moving inside, but nothing to be found. Also, the sound of utensils would be heard around the immediate area of the kitchen sink. One would say, maybe there were rodents in the apartment, but that was not the case.

One morning I woke up to find that that the wire on the phone adaptor was broken, during the process of my phone being charged overnight, and the immediate area was in disarray. Another incident was that a solid plastic handle on the transistor radio was totally cut in half - knowing for a fact that I do not sleepwalk at nights. When I got rid of the mezuzah, all unusual activities ceased to exist.

Before moving into a new home either your own, a rental house along with a yard space, you should be blessing the entire property including the yard space. It is the same as launching a sea vessel, a new craft or a new constructed bridge. In those times a ceremony would be taking place at the site. One should make a simple request or supplication to the Almighty God for protection from evil influence or principalities from high places and from the humankind intruders with bad intentions. Prior to resting in a room in Motels or Hotels, ask the Almighty God for safety, security and protection from evil spirits or any apparitions. This would give one a certitude, comfort, solidity and rest. People that had bad experiences with the above mentioned evil may sometimes become paranoia or fearful. So then, in order to cope with such evil, do not to be incongruous in evil- doings; or getting involved in any sort of iniquitous acts against God. Have a congregation of elders or spiritual leaders praying with that individual by either recommending or advising them appropriately in this matter.

Evil spirits or demons are subordinates to the devil or the Satan. Let it be known that their house is not divided; and that they are bounded on one accord. One should not attempt to find a cure from evil spirits or a cure from a curse by another evil, or in hiring a soothsayer or medium to cast out evil. It is considered an open invitation for more demons to enter the body and soul. An indissoluble remedy is to be spiritually strong and constantly praying by trusting in the Lord, so that no demons may be able to manifest themselves in your temple of peace.

A CHANGING ORDER IN THE COSMOS

The entire Cosmos is currently in a configurationally and spiritual evolutionary mode with drastic changes, which shall be occurring before the beginning of the end of the age in the year 6993 on the Prophetic calendar. Portent and iconic things will be witnessed in the sky. This is an authentication that the end of a 21 years moon cycle is about to happen. A theatrical action will be displayed in the Cosmos, and it may be viewed by scientists or armatures with their telescopes - also with a military or a commercial satellite. Prior to December 2013 the sun will be at its hottest in July 2013. This episode was in my vision while asleep, during the daytime.

Prior to the above mentioned event, something dramatic is going to take place in some place on earth in the year 5509/10. On that significant day, it was the crucifixion of Christ. The crucifixion had taken place on a new moon. Prior to that upcoming event, I had two separate visions of tidal waves coming into a city unknown to me. When God gave persons two dreams of the same, it will definitely be fulfilled. I had shared these visions with people close to me. In order to confirm those events that will be happening on October 2012, taking a look at the moon matrix of a 21 year cycle dated on the 13th day of the 1st month in the year 5509. That date is corresponding with the date on the Gregorian calendar on October 29th. Sometimes portent things are pre-destined to happen at some place or a spectacular things happening in the sky- and it is unstoppable. Many disastrous events are regarded as natural disasters. Those disasters may have a trend, and they are working in a cycle, and coming to pass after a period of many years- remembering that what is going around will be coming around. Also, upcoming will be the ending of a twenty- one year cycle that will be on the 15th of the twelfth month. Because of the variation of days between the Gregorian and Prophetic calendars, the time of event may be in 2014.

WARS:

It is not unprecedented to mention that world war two has phased out into world war three. Significant, but religious wars had soon started in the Middle East when a new nation had emerged. This new nation was given the name Israel in 1948. They had since occupied a portion of the treasured land, and legitimately claiming the land to be their own. A part of the Treasured Land was given to them by the United Nations in November 1947; that is two years after world war two had ended. It is now an ongoing problem with another group that is also making claims to the land. This is a prolonging problem and an unending dispute with no resolution in sight. This problem will not be considered a dispute, because the Geographic's of the Treasured Land shall be changed, prior the coming of the Messiah with a New Jerusalem.

A *one world order* shall be established on earth by the one and only God. *A new world order may be established by mankind.* Other religions may attempt to establish *a new world order*, but they may not succeed, because the people with the spirit of Lucifer are not in any way compatible with the population, that has the spirit of God within them. Because at the closing to the end of time, the people that are sincere to the Creator of the heaven and the earth- they will not be accepting *the mark of the beast*. Be aware, that the mark of the *Beast* can come in any shape or form, and also by trickery.

One should prepare themselves from such by sincerely making a supplication to the Almighty God for the faith and strength in resisting any mandatory law, forcing them to accept *the mark of the beast.*

The treasured land of the future shall not be named Israel, but will be called *'THE LORD IS THERE'*. A New Jerusalem, mountain high shall be relocated at the center of the earth, situating it's presence at the site of Dome of the Rock. Prior to that, the land shall be purged by fire.

The third world war which is in now in progress has been ongoing for a long period of time. It is being instigated by many extremists, tyrants and terrorists around the world. This is totally diverse from all previous wars. This war is unconventional, it is now spreading globally and it is unstoppable, because it is falling in the cycle. It is affecting most of the nations on earth financially, followed by innocent lives being taken- compounded with other unnecessary sufferings and homelessness. Some countries that are unaligned with the United Nations or the Western Nations are not absorbing the fullest impact and hardship, because of their non- affiliation with Western Nations. They are surviving by living off their farm land, along with their natural food and availability of resources. They are being identified as some of the poorest nations on earth. The names of those countries have never been heard of by many people, or their geographic location is known to most people in the world. Those scores of countries or Islands are not technologically literate and equipped with the latest inventions. There are over two hundred and forty countries on planet earth with most of them being unknown to the world.

Terrorism, tyranny, nuclear, cyber industrial espionage/warfare, famines, viruses, global economic chaos and instability of governments; followed by natural disasters and intensified tribulations are signs of awareness to the end of age. This is calculated on the Jubilee calendar to be at approximately 1,490 years away. However, Armageddon (Megiddo), the final battle will not be fought in the near future or in generations to come; as it is predicted by many writers and theologians. (The evidence is highlighted on the Jubilee calendar). At the end of these wars, man shall study no more war, but they shall destroy all weapons of mass destruction and sufferings.

REINCARNATION

Reincarnation is the rotation of souls on planet earth. All men must die, but their souls maybe returning to continue their unfinished mission on the earth. In some cases, a person may be pronounced dead by a medical physician. Their soul may reach to heaven just to be told that they have to return back to planet earth. Some evil souls are trapped on planet earth, but soon to enter new bodies of their kind. A typical place for the demons to find a new body to enter is at a funeral parlor – especially if that dead person was demon possessed. It is only the Creator of all things can give life to a body, but demons will only enter a living body, and not a dead one. God gives a spirit to everybody that is brought forth into this world by the woman. Those women that are deserving respect - it must be given to them by men or their children. God knows all things about that soul that he has created, because it came from His bastion.

The Prophetic life span of a person is 70 years, but if by any reason of them having the strength during their life span, they may live to be 80 years old; Psalm 90:10. However, a Prophetic 70 year is a Gregorian calendar 92 years. And the Prophetic 80 year is a Gregorian 105 years. Those souls that have passed on prior to 92 years or 105 years old may be sent back down to earth by God to enter a new body of their kind, so as to complete their earthly purpose or mission. That soul maintains records of memories in the past life, until the final resting place and judgment day. On judgment day Yahweh, His elders and His chosen ones shall judge the new resurrected body and souls of the wicked ones. This information given is not scientific, but it is spiritually motivated by the one and only Creator of the universes.

John 21:14 concurred that *Yehshua* had died, and within the days that he had spent on earth, He had adapted various human forms beyond the recognition of His followers. *Yehshua* did identify Himself to His followers by showing them certain signs that they could have recognized.

Matthew 17:10-13; 11:11-14. An authentic and classic case of reincarnation is that the question was asked by the disciples of *Yehshua* why Elijah must first come. *Yehshua* replied that Elijah had already come, but no one recognized him. Mark 9:13. *Yehshua* confirmed to the people that Elijah had already come, and he was John the Baptist.

Shem was one of Noah's sons. At the end of the age he shall be reincarnated and his spirit shall enter a new body. That body may be one of a prominent person in the Jewish community. Currently, the Jews worship and praise in the name of *Ha Shem*, which means in English, the name.

BORN AND DIE ON THE DATE OF BIRTH

Yehshua was born on the 14th day, at 9:00 pm in the evening of the first month, on the Prophetic calendar; and he died on the said day and time on Passover on a new moon. The only difference is that the various calendars are not working in conjunction with the Prophetic calendar of the Almighty God, which carries 480 days in a year and 40 days in a month.

Also, there is untimely death in this world, most likely soldiers at war, and those that were murdered, accidental death or they have committed suicide. The soul may not be placed into another body for a little while, until an appropriate body is selected by the Creator. That soul becomes a wanderer, or stay in a specific place of their choice. The spirit may be described as a friendly or unfriendly apparition called ghost.

All living souls will be judged at the end of times, while some may attempt to end their lives during the tribulation period, but without any success, Revelation 9:6; Jeremiah 8:3. Furthermore, painful tribulations are predestined for man. Bearing in mind, that a soul or spirit does not die, but it will live on through eternity, either to be tormented or live in peace. However, other religious organizations have their own dogma about that doctrine of the Christian church.

THE STARS

The stars were created separately on the same day with the sun. The moon has the same size disc as the sun, but the sun is seven times brighter than the moon. The measurements of the sun are attained by means of looking at both discs of the sun and the moon up in the sky during a clear day. The distance is measured, when both sun and moon are visible and opposite of each other in the sky. The moon that is seen westward in the sky is approximately at a distance of 480,000 miles away from the earth. An accurate measurement of the moon cannot be attained while it is in motion. Then, 480,000 miles may be the appropriate numbers in estimating the distance of the revolving moon, because it is consistently maintaining its distance, so as to obtain its light from the sun. The disc of the sun always seemed to be appearing at the same distance in the east at 480,000 miles, same as the moon; Isaiah 30:26. That distance is considered a myriad.

In Ezekiel 32:7, 8.God said that He shall take man and mankind off the earth. He shall cover the heaven and make the stars not shine. He shall cover the sun with a dark cloud; and the moon will be darkened in the end times. The question to the scientists is - what will the earth be revolving around, when there is no functioning sun or moon, that will be both stationary in the sky? The answer to that question is written in the scripture. God said that He shall be the light of the world. The sun light will be blocked out by dark clouds, so that it will not be shining its light.

THE SUN

A yolk of an egg represents the sun which comprises various organic matters, so as to sustain and expand the immediate universe, until the Creator destroys the heavens and darkening the sun.

Affirmatively, the sun is not one of a star, but it was created for a specific purpose. The sun gives its light to the moon when revolving around the heavens on its curvature circuit. The sun provides light unto the earth for a two twelve- hour periods in the day time in the Northern and Southern hemispheres. It is a conservative energy that whirls about continually in its circuit, and periodically it gets closer to earth at a 120 degrees angled south, which is directly in the middle of the summer as indicated on the 480 degrees figure "B" compass.

The sun which is a brighter light by day and the moon that is the lesser light by night in every 27 days, 28 days, 30 days or 29 days or less in a month of forty days on the Prophetic calendar. Job 38:24. By what way light separates itself which sends the east wind upon planet earth? Job 37:17. Does God quiet the earth by the south wind? Because out of the south comes the whirlwind and cold out of the north. Job37:9. God is in control of the weather, which is perfectly well with the animals, plants or insects. He can turn around the clouds upon His command, that they may do whatever He commands them upon the face of the earth. God causes it to come whether for correction or to His land for mercy. Job37:12, 13. *In Joshua 10:12-14 states that both sun and moon stood still in the battle of the Amorites. It means that both sun and moon were in motion, and they were not stationary in the heaven.* Mainstream scientists theorized that the earth is rotating around the sun. An accurate assessment is that the earth spins on its axis and shall not be removed, but only by the creator. Ecclesiastes 1:5. In the scripture verses Psalm 104:19, 22; Psalm 113:3; Psalm 19:4-6; all describes that the sun raises and the sun goes down. The sun provides light to the moon that it shepherd around the earth; and it provides light to the earth and all living organism. The sun is going forth on a daily basis, from the end of the heaven on its circuit unto the end of it; and there is nothing hid from its heat thereof. *Psalm 19:4, 6 states that in them He set a tabernacle for the sun, which is as a bridegroom coming out of his chamber, and rejoices as a strongman to run a race. His going forth is from the end of heaven and his circuit unto the end of it. There is nothing hid from the heat thereof.*

An equation in determining the quantity of electromagnetism (dust mite like atomic particles) throughout the universe is as follow:

The sun is revolving at 3,500,000 mph or 156464000 centimeters per second is a positive force, and the moon revolving at 500 mph is a negative force. The revolution of the sun, the moon around the Cosmos and the heavens (universes) in motion stabilizes the earth, moving it in a counter clockwise manner. Job 38:38: When do the dust increases into hardness and the clumps cleave together? The earth rotating at 960 mph or 42915.84 centimeters per second is a positive force that is known as gravity.

DELUSIONAL

An analogous explanation of delusion and a game change is that, when parked at a stoplight alongside a much larger vehicle than the one you are sitting in, and that vehicle slowly moves forward. There is a sense of feeling that your vehicle is actually in motion. An immediate response is that you step on the brakes. What the eyes are seeing at that instant, the mind becomes delusional. Another example of delusion is, when you are looking consistently at a mud wall or rock wall with water streaming down, the water will appear to be moving upwards instead of downwards and against gravity. This I had witnessed while walking a trail in the middle of the day in the very hot sun.

Planet earth may appear to be revolving, when actually it is rotating, because the heavens above are at the same time rotating counter clockwise along with planet earth, which is known as God's footstool.

In the winter season, the earth stays at angles of 30 degrees right, while the sun's rising and going down in the winter months is further away from the earth at 30 degrees. However, the moon's rising and going down remains on its said path, because the moon controls the seasons and the ocean tides. Genesis 1:16 states, that God has made two significant lights.

A CONJECTURE OF AN ELLIPTICAL GYRATION OF THE SUN

The sun is revolving around the heavens in the form of an egg- shape or elliptical manner and not similar to a rotating wheel, but it moves around the universe at a high rate of speed at approximately 3,500,000 miles an hour and 84,000,000 elliptical distance away from planet earth.

A Typical example is the same as having a red hot disc spinning around an object at a fixed distance. The results would be that the object will not be charred, because the disc is not remaining in a stationary position. By having a magnified glass with the rays of the sun steadily affixed over that object, it will start a fire or it will burn that object. If the sun is moving at a high rate of speed and at a safe distance away from the earth, it will not create a constant or a tremendous amount of heat constantly, but occasionally it will cause a forest or grass fire.

The sun travels on a separate circuit from the moon at the borderline of the heaven, which nothing is escaping the heat that it is generating. Psalm 19:6. When the sun goes down in the west of the earth travelling from the east, the setting of the sun displays a fiery color which varies in brightness in the west, depending on what part of the world it is setting. Theoretically, this color is reflecting off the red planet Mars when the rays of the sun are in line with the planet while beneath the earth. Actually a much lighter display of red is seen prior the rising of the sun in the easterly direction. The sun journeying to the north is circumventing the south which is in total darkness. The sun arrives from the north, travelling below the earth in line with the planet mars, going eastbound and down to the westerly direction during its twenty four hour period. In the south of the heaven where there is no sunlight, it is only thick darkness with endless space.

This is a mathematical simplification of the sun in motion: 84,000,000 ÷ 3,500,000 mph = 24 hours. These are some basic questions directing to the scientific community that is claiming the earth is revolving around the sun. If the sun's circumference is 2.7 million miles:

At what rate of speed would the earth be travelling so as to get to its starting and finishing point?

Does the moon simultaneously revolving around the sun in order to obtain its light?

If the cold winter is situated at the border of the earth and the earth is revolving around the sun, so then, where is the winter?

At what time period does the moon takes, in order to do a complete revolution around the sun?

The various scripture verses do coincides paradoxically that the sun rises and the sun goes down. The sunlight is reflecting on the moon during its gyration around the earth at an average 28 days every 40 days in a 480 days calendar year. Psalm 104: 5 states, that who lay the foundations of the earth that it should not be removed ever.

Job 37:17. Fair weather comes out of the north.

The heat of the actual sun does not falter, but it remains the same in both summer and the winter months. However, the sun is providing the appropriate heat in the summer months, but in the winter months, the earth is tilted to the left, and the sun will provide the necessary sunlight to planet earth. Progressively, the sun cools down the heated rays of the sun month by month, because the coldness or the winter at the border of the earth cools down the rays of the sun in the northern hemisphere. The sun then returns to direct east at the beginning of the seventh month in the summer, as the earth is retitled back to its original position.

There is no such thing as spring or autumn, only summer and winter which is divided at the border of planet earth. Starting from zero degrees in the first month, to 480 degrees a complete circle, the earth is however divided at 240 degrees, which is halfway to a complete circle. There is no right angle that is dissecting planet earth.

THE HEAVENLY DEMENSION

The Almighty God of the heavens and the earth cannot be reached by the human race, but only spiritually; or if beckoned spiritually by the Almighty for a visitation.

The heavenly space is stretched out like a curtain. Psalm 104:2. The gyration of the sun around the heaven and the earth in a year is 480 days. The moon revolution in the Cosmos in one year is 360 days. The ancient people used the days of the moon, so as to calculate the days in a month. That instrument they crafted was named a lunar calendar which carried 30 days in each month. However, the revolution of the moon in the Cosmos varies in a 21 year moon cycle. Therefore, the lunar calendar will have at times two moons in a month. In that case - is that extra moon not counted or is it considered two months in one month?

At the beginning of creation, there were seventeen moon revolutions in one year of 480 days with 40 days in each month. Thereafter, they continued to be three other of such moon cycle revolution in its twelve month period. That's in the 8th, the 12th and the 16th year on a 21year moon cycle. The Prophetic calendar of 480 days was used in ancient times by the prophets, so as to accurately predict an event that was given to them by God, and to be decreed to the human race. Somehow, that calendar to date cannot be traced. It was replaced with the lunar calendar or crafted by the Canaanites and subsequently passed down to the Babylonians as previously mentioned.

THE OTHER WORLDS

There is no dogma about the complexities of other existing worlds. The Almighty God who created all the worlds is in total control. He shall destroy them all, and create new ones to His likeness. Man's perception in the theatrical of other worlds is beyond comprehension to the human mind. The neighboring worlds are spiritual worlds that God had intended for the human race on planet earth to be dwelling on. The first two humans that He had created sinned against Him. This caused a transformation from angelic beings to mortal man. God created every spirit - good and evil. He has no beginning and He has no ending, because He is the first and He shall be the last. The other worlds are an unsolved mystery to the scientific community. None of mankind can venture into the other worlds, unless they have the mathematical formula, so as to open up the gates to those worlds or dimensions. Even so, the body of humankind was not created to endure the speed velocity in travelling into another dimension, only the spiritual being from the other planets can venture into the other worlds.

Hypothetically, when the moon completes its revolution around planet earth, it shoots outwardly into thick darkness in space in the northerly direction. The ten days absence of the moon reflects only on the Prophetic calendar. Job 38:19. Where is the way where light dwells and darkness dwells?

Planet earth is at 11,520 hours direct east (480 degrees or 480 days). In Job 38:18 God asked Job if he has perceived the breath of the earth.

The moon is at an average distance of 480,000 miles away from the earth and travelling at 500 mph. It is orbiting around the earth which has a circumference of 23,040 miles - while the earth is rotating at 960 mph.

Theoretical:

The revolutionary speed of the moon around the earth is 480,000 ÷ 960 = 500 mph. In a period of 30 days, the moon would cover a distance of: 500 mph × 24 hours × 30 days = 360,000 elliptical miles.

The moon's period away from the earth is ten days, because there is 40 days in a calendar month. Therefore, the mileage that the moon would cover in leaving its orbit around planet earth for ten days is 500 mph × 24 hours × 10 days = 120,000 miles. Therefore, the moon's total orbiting miles around earth and the Cosmos is 360,000 + 120,000 = 480,000 miles.

In one year of 480 days, the moon would have covered a distance of 480,000 × 480 days = 230,400,000 miles. In a twenty one year cycle, the moon would have covered 230,400,000 × 21 = 4,838,400,000 miles. In the beginning of the end of the age, which is the year of 6993 (333 full moons), the moon would have covered 4,838,400,000 × 6993 = 33,834,931,200,000 miles. So then, the recreation of planet earth in every 7000 years- the moon would have journeyed around the Cosmos and planet earth at a total of 33,834,931,200,000 miles.

The last moon cycle will always ends at 333 full moons, and the moon would be remaining stationary in the sky when there is a new recreation of the earth, and the sun shall be covered with dark clouds. The estimated years since the moon was created for planet earth is 47,510 years × 340 full moons (total full moons) = 16,153,400 cycle.

The moon is tilted at 1.92 degrees angle (960 mph ÷ 500mph) during its revolution around the earth. Its angle was designed, so that the surface of the moon will always face inwardly while gyrating around the earth. The moon's orbital miles around the earth, on the hour is 500 mph ÷ 24 hours = 20.83333333333333 miles. In one month the distance covered by the sun is 84,000,000 × 40 days = 3,360,000,000 elliptical miles. In a period of one year, it will be 3,360,000,000 × 480 days = 1,612,800,000,000 elliptical miles. In a twenty- one year cycle it will be 1,612,800,000,000 × 21 = 33,868,800,000,000 elliptical miles.

The total amount of miles that the sun would have covered at the end of the age are 33,868,800,000,000 × 6993 full moons = 236,844,518,400,000,000 elliptical miles.

Psalm 104:19 God appointed the moon for seasons; and the sun knows it's going down.

Psalm 104:5. Who lay the foundation of the earth that it should not be moved forever. Psalm 38:5. Who had laid the measures thereof if you know? Or who has stretched the line upon it? Job 38:6. Where upon are the foundations of the earth thereof are fastened? Or who lay the cornerstone of the earth?

When the sun and the moon have ended their revolution around the Cosmos in the year 6993, the sun would have completed 236,844,518,400,000,000 elliptical miles. The moon would have covered 33,834,931,200,000 elliptical miles. As mentioned above, the sun would have covered 33,868,800,000,000 elliptical miles and the moon at 33,834,931,200,000 elliptical miles, a difference of 33,868,800,000 elliptical miles.

Measurements are done through God's designs, His proper order and timings within the hundred and forty Jubilee years - bringing it to the end of every creation at the ending of a sabbatical year.

The seven remaining 7000 years are going to be seven plagues inflicted on mankind, with wars, intense tribulation and many deaths - with the dead coming back to life. The tribes of Israel shall be returning to the Treasured land of Israel. The book of life shall be read, and those that are in the book shall live. All sinners will be cast into the lake of fire to be permanently enduring sufferings by the everlasting fire, along with the Satan. 2 Peter 3:10 states that the heavens shall pass away with a great noise and the elements shall melt with the fervent heat. The earth also, and the works that are therein shall be burned up. Psalm 102:25, 26; Isaiah 51:6; Revelation 20:11. God said that He will shake the heavens; and the earth shall remove out of place. *Isaiah 13:13; 34:4.*

The Almighty God who is the Creator of the heavens and the earth has created seven heavens or seven other distinctive worlds. However, planet earth is His choice in dwelling among His people in the new world to come.

The seven spirits of God travel on its clouds to all the heavens which are located at the right hand or the North of the seventh heaven, a dwelling place of God. The heavens were created in six earthly days, earthly time; and six thousand years in God's time-line.

Looking at figure B, the 480 degrees compass, it indicates that creation in six days had ended in the north (360 degrees or 8,640 hours).

The Real Estate of the seven heavens - including the one that will be destroyed is approximately 161,280,000 square miles. The heavens moves along the vast empty and endless space coupled with dark clouds. The heavens are in motion at an average speed of 960,000 mph. They move at 6,720 elliptical miles every hour (161,280,000 ÷24,000 mph).

ELECTROMAGNETIC FORCES

The force of a gravitational pull, coupled with electromagnetism is generated by the sun and other worlds. The moon's negative force during its revolution around the

Cosmos, with the rotating worlds helps to stabilize the earth on its axis; they all cause the earth to rotate at 960 miles an hour. The earth is consistently covering forty miles every hour. It takes the earth 1/480 degrees through a turning angle each day and 1000/480,000 the heavenly motion every one thousand years.

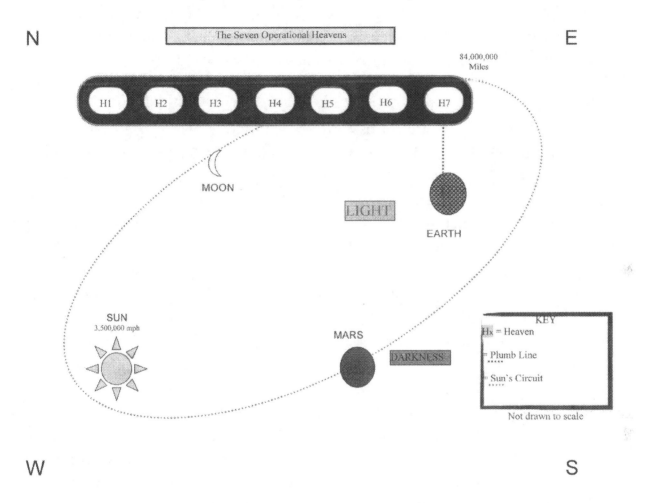

Joshua 10:13. The sun stood still in the heaven and it did not go down.

Electromagnetic forces are perpetually interacting with nature in general, even animals, the human race; the birds of the air, the fishes in the sea are somewhat magnetized or connected with the electricity in the universes. These forces are exceptionally having a sustaining effect on every living organism; and at the same time it is keeping afloat all physical phenomena in the other worlds.

Theoretically, the electromagnetic forces are fundamentally controlled by a vast universal magnetic field that is located in the northerly direction at 360 degrees. In accordance with a 480 degrees magnetic compass, each worldly dimension or direction is calculated along the line at 8,640 hours out of a complete circle of 480 degrees - or to the final hours at 11,520 hours in that circle. The cardinal points in degrees are east at 480 degrees, west at 240 degrees, North at 360 degrees and south at 120 degrees.

The concept of a unified physical power is controlling all magnetic forces that are being generated in the worlds. These powerful forces cause the circular momentum of the other worlds and everything in space to move in a synchronized rhythm. This unique magnetic field is identified as a powerful force which is situated in the north at 360 degrees, and 120 degrees left directly east, which is at 480 degrees.

WHIRLWINDS

Cyclically, whirlwinds are consistently and constantly being generated, because of the whirling motions created by the seven worlds, along with the velocity of the sun travelling at 3,500,000 mph on its circuit. The whirling motions are working in conjunction and the timing in a 21 year cycle - what goes around comes around. In Job chapter 37:9 it states that the whirlwind comes out from the south and the cold out of the north. Job 37:17 He quiets the earth by the south winds.

A 21 year cycle has a chart that is accurately forecasting all propensities or a trend to follow at a certain time line. Mother Nature is working within the frameworks of this chart. Consequently what is naturally exhibited in the universes or in the Cosmos and on planet earth is generated by the forces - and it is directed by the Almighty God.

Things may appear to the scientific community, that those forces in the universe and in the Cosmos are being functionally and mechanically controlled by its own force. That theoretical assessment is inaccurate, because any mechanical device at some point and time had to be created by someone or a being. That spiritual being is the Almighty God who created everything - and who is totally in control of what He has created and still creating. In examining that all mechanical devices will at some point and time malfunction, so then it will have to be maintained, or some sort of repairs have to be done on that device. In that case, who so ever created that device will have knowledge in doing repairs and maintenance to that device - so is the God that created all things. He is the author, the planner, engineer, architect, mathematician and doctor in all matters pertaining to His creation.

MEASURING THE DISTANT SUN

Ironically, no accurate measurements can be taken from the sun which is always in a revolving motion, twenty four hours seven days a week. It is incomprehensible for any of the human race or humankind on earth to take measurements, because measuring the sun can be done only mathematically. Furthermore the sun rides on its circuit twenty four hours seven days a week, only around the seventh heaven and the earth - but it does not venture into other heavenly spaces or dimensions.

When looking at a distant sun with the naked eyes, one will not be able to follow its movements, because the sun is moving clockwise, and simultaneously the earth is rotating counter clockwise. Therefore, the eyes would believe that both objects are motionless. It is considered a delusional moment, even if one is looking through a telescope.

Typical example:

When the body is spun around at a fast rate of speed, the equilibrium which was created within the body of every living species on earth would cause an imbalance and disorientation, same as when a human is very old in age. The reason being, that the spinning object will be out of synchronization with the rhythm of the rotating earth.

The sun and the moon are like a crux- disc that is magnetized around the Cosmos in their respective circuits on a timely basis.

7000 YEARS RECREATION OF PLANET EARTH

Every seven thousand years is the recycling for the earth's purification and the heavens in every 7,000,000 years, God shall be purging the earth with fire and recreating it in the year 6999/7,000. The age of the earth is now in the year 5510 according to the Prophetic calendar. *At the end of times there shall be no new moon on that day or after, because its last brightness shall shine on the earth in the year 6993- the 333 moon cycle (6993 ÷ 21 years= 333). The Jubilee calendar carries 140 Jubilee years. (140 by50 =7,000 years).*

The moon will not be revolving in the Cosmos, and the sun light will be blocked out by thick dark clouds. The stars in the heaven shall fall to the earth, and everything in the heavens, including its hosts will be affected during the recreating of planet earth and the heavens. Remembering that a Jubilee year on planet earth is 50years and in the heavens are 50,000 years – One earth day on earth is one thousand heavenly years.

COMPARISON TO A CENTRALIZED SUN

Scientists from the fifteenth to the seventeenth centuries in their farfetched theory claimed that the earth, moon, stars and other planets are revolving around the sun in the universe. Science text books, magazines and journals educated the mass that the earth goes around the sun - this is very cogent to the scientist. If that's the case, can the scientific community say what distance the earth is being maintained away from the sun at all times, so as to have winter and summer? If that's so, the following should be taking place, because the earth is not protected within a glass proof bubble, but somewhat protected by a thin ozone layer. So the following will definitely permeate the earth's environment: Radiation contamination; frequent collision with outer space junk; asteroids; hydrogen; methane; helium gases; solar dust; etc. These obstacles will cause the earth's ozone layer to mal- function. The clouds, along with the chem. Trails would not be able to withstand the speeding velocity that the earth would be travelling around the sun at approximately 67,000 mph, because everything around the earth would be disintegrated or contaminated. The prognosis will be that the bubble will eventually burst by the high speed velocity of earth revolving at 67,000 miles per hour around the sun.

STARS

The stars in the heaven are countless, and no man is able to calculate the amount of stars in the sky, because some expires while some are born. Genesis15:5; 26:4. When stars expire, they display a fiery trail behind them during their pitch. Stars are somewhat made up with a compound of gasses, that subsequently they are burnt out when they expire, and those visible gasses are left out in the vast space.

To reiterate the fact that there are more heavens: *Psalm 8:3- consider my heavens that they were my fingers - the moon and the stars which I have ordained.* Without a moon and its magnetic force, the earth will not function as it should, because the moon controls the two seasons. In Psalm 104:5 Yahweh said, that who lay the foundations of the earth that it should not be removed forever. This statement is absolutely unpretentious and immovable.

In the beginning of the age at five thousand, five hundred and ten years ago, there was total darkness upon the surface of the earth. By the word of God, heaven and earth came into existence; and God separating the light from the thick darkness that was intermingled. Genesis 1:3-5; Psalm 115:16. God made a path similar to a circuit or highway to accommodate the sun and moon obstacle free, so as to have them revolving around the galaxy or earth respectively; and to give light to the earth without being blocked by any foreign objects. *If there is no designated circuit, no sunlight would be able to penetrate through the thick darkness and many obstacles in the Cosmos, so that it may shine on planet earth.* Only the clouds around planet earth would be able to block out the rays of the sun and the light of the moon.

THE MOON

The moon shines, because its surface maybe composed with a type of dysprosium or aluminum metallic element acting as a reflector. Those elements are reflecting when the sun light beams on it.

The lunar (moon) calendar also needs reforming, and should not be used as an agricultural almanac, because it is highly inaccurate. New or full moons appearances to the earth are not on a consistent basis in the twelve months in the year. The lunar calendar dates for sowing and reaping will go adrift from time to time in the two seasons. If the present calendar system remains, then all the agricultural industries worldwide shall falter. Nations will suffer financial losses and food shortages will be increasing tremendously and food prices shall escalate.

The moon has its own designated circuit that is extending around the Cosmos. Its highway is much shorter than the highway of the sun. Logically, it is because of the distance away from earth the moon appears not to be moving, but standing still. The reason is, because the moon is revolving clockwise while planet earth is rotating counterclockwise; and it balances off the movement of the eyes.

The earth's total hours in rotating for a year at 480 days is 11,520 hours.
Below is a countdown as to how the earth is rotating:
960 mph x 12 = 11,520 hours (480 degrees/ days)
960 mph x 11 = 10,560 hours (440 degrees/days)
960 mph x 10 = 9,600 hours (400 degrees/days)
960 mph x 9 = 8,640 hours (360 degrees/days)
960 mph x 8 = 7,680 hours (320 degrees/days)
960 mph x 7 = 6,720 hours (280 degrees/ days)
960 mph x 6 = 5,760 hours (240 degrees/days)
960 mph x 5 = 4,800 hours (200 degrees/days)
960 mph x 4 = 3,840 hours (160 degrees/ days)
960 mph x 3 = 2,880 hours (120 degrees/ days)
960 mph x 2 = 1,920 hours (80 degrees/ days)
960 mph x 1 = 960 hours (40 degrees/ days)

VANTAGE POINT

Operating from a vantage point at JFK Airport New York City at 8:30am when the full moon is visible on a clear morning, it is always appearing along with the sun high in the sky in the westerly direction. When the full moon is appearing it is seen rising in the East, and its appearance is very large and looking closer to earth. That scenery is not consistent, because the calendar days are not the same. However the moon goes down in the west, prior to the sun visibly appearing in the east. In some instances the moon is seen in daylight in the sky.

The moon has a negative gravitational force generating from its disc, while the earth carries a positive gravitational pull. What belongs to the earth clings to the earth, because of its magnetic field. Every living organism contains a small amount of electro-magnetism that causes them to cling to the earth.

At a vantage point in visualizing the movements of the moon around the earth in the months of August and September of 2009, measurements were taken with a transparent rule in the vicinity of JFK Airport New York City, USA:

August 30th   *- 08:30 pm -   SSE- 10 inches - right below a star*
August 31st   *- 08:30 pm -   SSE- 10 inches - right below a star.*
September 1st   *- 08:30 pm -   SSE- 1.5 inches- high right above a star.*
                   - 12:30 am -   SW- 1.5 inches- lower of star.
September 2nd *- 08:30 pm -   SSE – 2 inches- lower left of star.*
                   - 12:30 am -   SW – 3 inches – higher left of star.
September 3rd *- 08:30 pm -   SSE- 2 inches- lower left of star.*
September 4th *- 08:30 pm -   SSE - 6 inches- lower left of star*
September 5th *- 08:30 pm -   SSE – 9 inches – lower left of star*

Clearly it is recognizable that the star mentioned above is a stationary star known as Polaris or the North star in the heaven, because it was used as a point of measurement between the locations of the moon at various times. This proves that the earth is tilted at an angle of 30 degrees during its rotation, so that when the full moon is in the west, the sun then appears to be rising in the east visa versa. The earth however rotates counter- clockwise.

The moon couples with the earth throughout the winter months and the summer months, because it is controlling the seasons. In the beginning of winter, the sun appears to be moving further away south of the earth for a period of 240 days, and then reposition itself in the starting of summer for a time of 240 days. This is not the case of the sun. It is the earth tilting that makes it appearing, that the sun has moved further away from the earth.

In the middle of summer, the sun is at 120 degrees overhead planet earth. It is considered the peak of the summer season.

The moon is full in three consecutive days:

- *Full with lesser light.*
- *Full with a brighter light.*
- *Full with a dimmer light.*

There is only one moon in the immediate seventh heaven that God created over five thousand five hundred and nine years ago. The said moon continues to be revolving around the Cosmos in a period of 28 days at times in a time period of 40 days. The moon, on its completion in revolving around the earth will head to the northerly direction as indicated on the moon matrix. Then it will return at approximately 28days, a one- time 30 days and one- time 27 days in the year. To reiterate, the sun is working in concert with the moon, as it is revolving in an outer circuit, while the moon is revolving closer to the earth on a shorter circuit, but clockwise with the sun.

CRITICAL ANALISIS

Some pertinent questions to ask some of the mainstream scientists that are abnegating that there is no God who created all things, but claimed that the universe was created by itself. And the scientists who are believers in the God, are not speaking out to tell the whole truth about their concept of creation:

- Who created all man and mankind?
- *Who gave them a spirit in their bodies? Ecclesiastes 3:21; Isaiah 45:12; 1 Corinthians 2:12.*
- Who gave them a mind to make choices of knowing well or doing evil?
- Who takes away the soul from the body?
- Does man has the power over the spirit to retain it?
- Can a scientist who is not a believer cast out demons from a human?
- Can a scientist walk on water or calm a storm?
- Can scientists verify the existence of the universe or nature?
  At the end of the age God shall conceal the sun with a cloud, and if that is so, why should the earth revolves around a sun that would be voided? God will be the light of the new world to come. Isaiah 60:19, 20
- Can scientists cause a woman to reproduce without the male sperm?
- Can scientists reproduce the various type of flesh of animals, birds, reptiles, fishes and mammals? Negating that there are no parallel planets like the earth in our immediate world. God had created the earth especially for Himself and created man in His image, so that He could dwell among His chosen ones. However, blasphemy was committed in the heaven and the devil was banished with his contingent of a third of angels to the only earth-like planet in the immediate universe or the seventh heaven. The earth is connected to the seventh heaven by a plumb-line.

God's word is the truth and nothing can substitute or replace any concept or precepts of His creation. Maybe, they are asking that who is the almighty God that they should honor Him; and what benefit could them achieve from Him? The fact of the matter is, that some scientists becomes agitated or annoyed when told that a spiritual being has created all things in the universes and planet earth.

They are the ungodly ones that continuously prosper in the wealth of the world, because they do not have the spirit of God within them; and they show no appreciation. Psalm 115:16 states, that heaven and the heavens are the Lord's, but the earth has He given to the children of man.

Religion is ones belief or a community affair that should be giving back to the poor, orphans or needy. The religious organizations duty is to support law enforcements, visit residents in prison; visit the hospital institutions, and should be spiritually motivating the troubled youths, instead of arousing ones emotions for personal gratification and benefits. The younger generation should be striving to become an asset in their community; and to work much harder to be inspirational to their peers.

Psalm 68:33 states, to Him that rides upon the heavens which were of old; meaning that the previous heavens were there prior to the new heavens that He had created five thousand five hundred and nine years ago. However, mainstream scientist are seeking new innovations, creating new methods for a better life for humanity, plants and animals, by working under difficult conditions and with limited resources.

Those unanimously recognized scientists achieve professional excellence in various fields of science on an annual basis. They are enhancing better quality of synergistic pharmaceutical drugs to prolong a person's life. However, the monies go into the pharmaceutical corporations that are manufacturing drugs, not for curing patients on a short term basis when it can, but the drugs are made, so that treatment will be given to the patient over a long period of time. In a whole, scientists are the visionaries that possess great amount of talents, dedication, skills, somewhat accurate at times over the past four decades, because of new innovations and new invented technology, so as to enhance their work.

In the book of Job 36:24-26 it stated, that we all magnify his glorious work which is seen by man. We behold it from afar off, but we do not know Him, nor can we determine the number of years, even though some men do not acknowledge, that there is a creator who created all things; and made everything possible for man and mankind to be enjoying the beauty of His creation. However, God is continuing blessing them with talent, wisdom and wealth. Mankind should therefore comprehend that God's word is a proclamation with the greatest power ever; and everything adheres to His the commands in-which He is constantly executing.

STRESS

The world that our previous generations enjoyed, were with less technology, less stress, a better life, but they are not the same in this present generation. The younger generations in the world that we occupy are not so sure what the future holds for them. Many pundits in the various fields of science, economics, along with the other mainstream Medias give a very bleak hope of life for the future of man and mankind on this planet. It may be the reason why some youths are becoming more reckless with their lives, violent, carefree and unmanageable in the schools and at home.

THE QUALITY OF FOOD, AIR AND WATER

The older generations can be a witness to the drastic and rapid changing in the food qualities that those corporations and companies are producing. They do not have the required nutritional value for human consumption, but the food has to be backed up with supplements, so that one can remain healthy and live a long life; and also by properly maintaining their immune system.

The majority of food products on the shelves which are being displayed in the groceries on the shelves contain genetic modified organisms [GMO]. To that effect it is reducing the quality of life for the human race by shafting the present generation with low grade quality food of various kinds. Maybe there is no alternative, because of the increasing population in the world.

The farming industries and the multi-corporations intension is not to disclose the true facts or ingredients in a product, because most of the consumers would not buy the food item in the food market. Shoppers are willing to spend a little more money in purchasing organic food products for the benefit of their health. Auspiciously, the life span of previous generations that are alive today has significantly increased due to 100 percent food quality that they had in their generation.

The taste or smell of fruits and vegetables in the market places cannot be compared to what was grown two generations ago. The various type of meats consumed today are of a low nutritional value to the human body. Some are not appearing to be organically grown on farms, but in some instances look as if they were dyed with artificial coloring; or enhanced and puffed up with additives.

Over the past centuries the air quality in the developing nations has diminished due to new innovative technology, nuclear plants, chemical factories, exposure to dirty electricity, unsecured land filled sites and an increase of chem. Trails. Those remaining trails in the skies are appearing to look similar to the regular clouds. The drinking water and water to irrigate the land in the undeveloped countries is becoming disastrous to those nations; and the lands are becoming barren or due to the lack of rainfall and digging of deep water wells.

HYPOTHESES:

This is a simple theorized analogy of an interpretation as to how the universe was formed at the beginning of time. The below constructed, but creative presentations display two dimensional phases; a spiritual and a physical.

As illustrated below, it is a formidable perspective and an inescapable condensed darkness of a universe that we are contained in. It is made up of various elements, energy, matters which are originated by the one and an only almighty God who is surrounded by thick dark clouds. Psalm 97:2.

Many people ask which comes first – the chicken or the egg. The answer to that question is that both egg and chicken are one and they come together. The hen lays the egg which comes out very soft, before it touches the ground and becoming hard. Inside of the shell casing there is a chicken in a primitive form that is attached to the yolk of the egg. This example is similar to the precept of the time in the creation of the worlds.

The precepts are methodical and dimensional, which is a physical outlook by using a cracked egg and placing the yolk of the egg with the white portion on a flat surface.

The yolk is representing matters such as: helium; oxygen; hoar- frost; nitrogen; methane; hydrogen; dust; darkness; light; rocks; also, other elements and other gasses. The white of an egg represents space without any measurements attained, or obstacles floating in space. Psalms 147:15-18; 148:5-12; 135:7.

Spiritual Precepts: God – His hosts - chosen souls - Lucifer - space – the word - command executed – microwave- matter- explosion – dispersion – curtain like design – other heavens or worlds – the earth as the corner stone. The end product of God's creation, as per His specifications on his blue print looked like a curtain spread out in the heavens. Psalm 104:2; Psalm 33:6. The fact of the matter is that the universe was created by fire and it will end with fire. A fire cannot be ignited without a spark and without the presence of oxygen. So then, the Almighty God in His magnificent power was the only source in creating the heavens and the earth along with every particle in them.

To reiterate the continuous refashioning of planets and stars in the universe is like a similitude to an uncooked yolk of an egg on a flat surface. An accurate assessment in the process of forming into a round shape is that extreme heat molds the yolk of the egg in a hard round shape when it is intact and boiled with the shell. The same concept will be taking place when God shall be creating, recreating or expanding His universe (heavens) by fire. The destruction of the universe will be similar as placing a hard- boiled egg in an active microwave oven for a few minutes. The end result will be that the heating of the microwave will disintegrate the egg and scatter various sizes of particle all over the oven.

ENTROPY:

To reiterate, God's strategic designed plans on a compilation of heavens were to be a whole body and to dwell among His creation. However, God's ordained angel Lucifer had planned a take- over (coup de tat), because of jealousy, greed and a power struggle - the same as man and mankind on earth. The outcome of their punishment is that he and a contingent of angels were banished to be a dominant force on planet earth, until the end of the age in the year 6999, and to face a final judgment. Entropy shall continue to exist in heaven until God's destruction of the universe and His recreation of new heavens and a new earth. Man was given the opportunity to choose, either to become righteous or choosing evil, but man chooses the latter, which resulted in a woman being beguiled.

- Job 3:18- God put no trust in His servants, and angels He charged with folly.
- Job 15:15- God put no trust in His saints. The heavens are not clean in His sight.
- Job 25:5- Even to the moon, and it shines not. The stars are not pure in His sight.

LIFE SUSTENANCE:

- To summarize, the sun is representing a yolk of an egg that comprises all matters to sustain life in the galaxy. Without the sun there shall be no life existence on earth, along with the stars in the heaven or the other planets until God returning to dwell on earth. Beneath the earth life is in existence, because of its fire that is rotating and providing heat to the humankind. Job 28:5.

- The sun, the moon and other hosts of heaven are God's ingenious creations, and they should not be bowed down to, because they are not the creator of things, but His creation. *Psalm 84:11 states, that God is a sun and shield giving grace and glory to the upright.* Man and mankind cannot comprehend or measure the time of creation or the years of the Creator's existence, because the current Gregorian calendar that is in use is totally incorrect and fallacious.

- The earth was created at the beginning of time, recreated many other times and continues to be recreated according to the scriptures, so time actually cannot be measured. The earth and the heavens have a cycle that is working within a time frame along with the Jubilee calendar. They have the same concept, but the years are lesser on planet earth by 480,000 days in heaven, in comparison to one day on earth. One of God's 7 spirits dwelled on earth before Abraham was born. He was the king of *Shalem* and the high priest in the Order of *Melahszte* that blessed Abram after his defeat of the five kings.

- In Matthew 24: 22, it states, that the mercy of God may shorten the days with the terrible afflictions destined for man and mankind on the earth. But shortening of the days may happen, because of His chosen one's sake. Mark 13:19, 20. God shall dispatch His host of angels to gather together His chosen ones from all corners of the earth and the heavens. Mark 13:27. Remember, that the last enemy to be destroyed is death. 1Corinthians 15:26; Revelations 21:4. He whosoever loves his life shall lose it; and he that hates his life will keep it. John 12:25.

AFFIRMATION

The modus operandi of the creator of the universe is for Him to be back on earth. God's choice is to live among His chosen angelic like ones instead of angels. God shall live in peace and joy with His people who shall worship Him in truth and righteousness, without evil intermingling among His chosen ones. God said that every knee shall bow to Him and every tongue shall confess or swear to Him. Isaiah45:23; Romans 14:11; Psalm 63:11.

The secret to wisdom, knowledge and understanding of the worlds, in becoming a good and well - rounded person is to study the word of God.

CHAPTER FOUR

FLUCTUATING WEATHER CONDITIONS

Climate change is not a reality, but it is the man made Gregorian calendar that's misleading the climatologists and meteorologists in their forecasting of the weather. Over the past centuries they have been losing their ability in predicting accurate weather forecasts- both on a long and a short term basis due to the weather frequent fluctuations.

The winter and summer seasons are assigned to the borders of the earth, and they are working in conjunction with the planet when utilizing the Prophetic calendar of four hundred and eighty days in the year. The world is not fully aware that the Gregorian calendar is shorting one hundred and fifteen days in the year. However, the effect is being felt especially by farmers on a year to year basis in both seasons; and through their inaccurate weather reporting on an annual basis.

-   Ecclesiastes 3:1 states that to everything there is a season.
-   *In Psalm 104:19, it states that Yahweh appointed the moon for seasons and the sun for its setting.*
-   *Genesis 1:14, Let the night and day be signs for seasons, also the days and years.*
-   *Psalm 74:17, God has prepared all the borders of the earth. He has made summer and winter.*
-   *Job 14:5. Seeing that the days of God are resolved, the number months are with Him.*
-   *Job 17:12. They changed the night into day and the light is short, because of darkness (Day light saving time)*
-   *Deuteronomy 17:3. They have gone and served other God's; worshipped the sun, moon and the host in the heaven which God had not commanded them to do.*
-   *Psalm 136:8-9; Genesis 1:16. The sun is to rule by day, the moon and stars to rule*

in the night-time. None of the above scripture verses has indicated that man should create a calendar by the rotation of the moon or the sunshine by day.

- *Isaiah 38:8, a sundial of King Ahab was turned 10° backward by Yahweh, so that the sun returns to its original position.*

A FALSE PREMISE

Based on the above mentioned scripture verses, the scientific community has no substantial evidence to prove, that false premise of a critical climate change or Global warming that is taking place on planet earth. They are claiming that Carbon Dioxide which is being exhausted from vehicles and factories, or the burning of coal along with other green- house effect is contributing to the changing in the weather. Geo-engineering is also causing a change in the weather. The scientific community did not say that the chemicals being sprayed in the air on a daily basis can partly be responsible for weather change.

However, Carbon Dioxide is very vital to planet earth, because the gasses normally form a canopy over the earth in order to contain its heat. The cloud of gasses that is released from an erupted volcano plays an important role in maintaining the ozone layer. The hype in the inconsistency of weather change shall continue, unless a current calendar reform is flitted and replaced with the Prophetic calendar - a calendar that is working in conjunction with the summer and winter seasons.

There is a misconception or misunderstanding that there are two seasons, and not four seasons on planet earth. *God assigned summer and winter at the border of the earth.* The sun and the moon are totally controlling the two seasons. When the sun is directly overhead the earth, it is at 120 degrees - as per the 480 degrees compass. The rays of the sun has a full impact on earth, and it makes the heat more intense in the mid- summer.

Slowly, the sun is appearing to be getting closer to earth at the beginning of the first month on the Prophetic calendar. After proceeding to the end of the sixth month, the sun seems to be gradually moving further away to the right of planet earth, but it is not. It is actually the earth that is tilting to the left at approximately at 30 degrees. This is the beginning of the winter months in the seventh month on the Prophetic calendar.

On the Gregorian calendar it is conceived as three months of spring. The positioning of the earth is now at 240 degrees or completing 240 days or 5,760 hours rotating on its axis during the gyration of the sun.

In an experiment with a steel needle in a container of water it was observed, that on October 10th 2012 on the Gregorian calendar, the needle began moving to the left. Finally, it stopped at 330 degrees, a decrease of 30 degrees from 360 degrees north.. It was observed that the sun was not directly overhead. It has also been observed that when the sun was going down in the west, it was angled at 30 degrees to the left of the earth. Its ultimate purpose for tilting at 30 degrees angle is that winter is located in the northerly direction at the border of planet earth. The sun had appeared to have changed its course, but it did not.

The sun remains on its super highway at a high velocity of speed in twenty four hours and seven days a week throughout the year of 480 days, and millions of miles away from the earth. The naked eyes would not be able to track the movements of the sun's swift motion, because of its velocity at 3,500,000 miles an hour. The sun creates solar winds, and the earth rotates in a counter clockwise manner.

The vast blanketed universe had been created to neutralize, acclimate and evaporate environmental hazards or the toxic vapors from industrial complexes on earth. The Forest fires, the nuclear clouds, meteor dust and other burnt out hosts of the heavens are vaporized when it is close in the path of the sun. They can also be left out into the massive space as condensed fog- like appearance, either dark or cotton- like material, that is, when they are further away from the sun, as indicated in many science magazines or the internet.

OZONE LAYER:

The leaves of the plants and trees absorb Carbon Dioxide from the air in order to flourish and for their survival on earth. The ozone rainbow type layer immensely protects the earth from any foreign gases and the sun's ultraviolet radiation. It functions in a way similar to a circular valve whirling around the earth, and acting as a flue releasing unwanted gasses in the vast universe. If there is a constant depletion in the ozone layer, the earth can be affected tremendously by ultra violet rays of the sun. More so, people will become cancerous or darker, and there will be a high increase in forest fires and elsewhere. That is, if the sun is in a stationary position and not revolving around the earth. In that case, the sun will be definitely destroying every living organism on planet earth, either by drying up the oceans, causing the planet in becoming barren as other planets, or the land will definitely become charred.

WEATHER INCONSISTENCY:

The previous generations, that one may have recalled is that the cold season normally begins the day after Labor Day, in the month of September in New York City. Other years, the cold season had started in the month of November or December, and about ten years after, it goes back in the month of September. It shows an inconsistency in the weather pattern. Its fluctuation on a yearly basis is due to the drifting Gregorian calendar that the nations of the world entrusted in giving perfect weather predictions. The Weather Forecasting System shall always be unpredictable, because the Gregorian calendar is shorting days in the year. It is going adrift as a ship without a serviceable engine or a sail when it is not anchored at sea. The calendar will continue misleading mainstream scientists that global warming is at our door steps.

CHANGES:

The refashioning of land contours or landscapes are mainly caused by frequent earthquakes that planet earth is experiencing in the latter years of developing. The melting of glaciers in the Arctic region was always partly a defacement of the earth in its expanding mode and progression, same as man, the animals, birds and plants, that are going through changes in their physical structure. Even though, some scientists who are theorizing climate change and global warming, may not have documented weather data in past centuries ago. They are lately acknowledging and becoming alarmed that changes are taking place with the weather pattern; and they are drawing to a conclusion that the earth is over- heating. A logical assessment of disastrous claims that are being made by mainstream scientist is indecorously, because the changes happening are in accordance with mother- nature.

The scripture gives relevant information and the history of changes as to what the earth was before, and what it is now experiencing, but it is the same cycle all the time. Frequent earthquakes had been taking place in the past and throughout biblical times. The shifting of the earth or submerging land in the sea, sinking of lands under water, new lakes and many volcanic eruptions is partly the natural or expounding process of mother earth. The entire Cosmos is appearing to be going through a normal cycle of changes, which will be continuing until the end of days. In scriptures, it describe that Mount Olive shall be splitting in two, and it will be creating a gorge. A process in the gradual changes will be continuing at the end of a twenty- one year cycle on the 15th day in the twelfth month on the Prophetic calendar in the year 5509.

The younger generation of scientists, were not present in order to be recording or witnessing these changes in the past hundreds of years ago. Those claims are based on their predictions on global heating of planet earth. If only scientists could have checked in the archives of the scripture, so as to obtain relevant information about the earth and its progression of changes, along with the frequent earthquakes; then they would have taken a different approach and conceptualized things differently at what are taking place on planet earth. Many of scientists are not believers of a God that created the vast worlds. In *Corinthians 14:33 states that God is not the author of confusion.*

In accepting and adapting the rebirth of a Prophetic calendar as a World's calendar, all nations on earth may benefit with the following: a better economy; having new innovative possibilities and solutions; accurate predictions and a resolution to the fact that there is no climate change or global warming. *Proverbs 26:1 states, that as snow falls in the summer, and as rain in harvest, so honor is not seemly for a fool. Proverbs 25:13 states that as the cold of snow in the time of harvest, so is a faithful messenger to them that sent him.*

CONVERSION:

In reducing greenhouse gas emission by switching to alternative energy or the conversion to land, air and maritime vehicles to hybrids, solar energy along with other conversions are not solutions to a non- existing problem that is being created by mankind and his calendar.

ILLEGAL DUMPING:

Dumping of illegal waste matters into the sea and water ways, instead of burying the wastes on land, contributes to environmental hazards. The increase dumping of industrial waste matters into the sea by other nations can be partly contributing to the rising sea tides. New additional policies, protocols and laws should be in place, in order to minimize such dumping of industrial waste matters into all water ways. Dumping of waste matters contaminates the fishes of the sea, other marine life, and it may be affecting the birds in the air and the human race, that are obtaining their daily food from the sea. However, the United States of America could be vouch for that their nuclear wastes and other industrial waste matters are properly stored away in safe containers, or buried on land and away from inhabited neighborhoods. The earth's populace is consuming nuclear waste matters and other unwanted chemical matters into their system.

If the situation is not taken seriously or rectified as soon as the earliest, and laws are not put into place, then unnecessary monies would have to be spent medically treating the world's population or exhausting monies to that effect.

STANDARD CLIMATE:

The good news are, there is no global warming or change in the climate, not to mention that thousands of air buses, jet planes or aircrafts are spraying chemical in the air on a daily basis in some states in the USA. They can be one of the main contributors to environmental hazards to life on earth. UFO's, meteorites and other junk that penetrates the ozone layer by hitting the earth, should be considered pollutants to a non existing problem claimed by mainstream scientists. This is not the case, because dusts and rocks clumped together to form planet earth at the time of creation.

The climate conditions would not be changing unless man or mankind ceases in attempting to modify the weather by spraying the skies daily, such as with aluminum and other chemicals. A reasonable question to ask the scientific community about carbon dioxide that eight billion people and animals are exhaling - is it affecting the ozone layer? The God of all worlds had designed a plan for the changes in His seven heavens and planet earth. God is expecting planet earth to be running out of land space for the burial of waste matters and the dead. He had already envisaged that the earth - *His footstool,* would become polluted and over populated, because God did not create the earth to be populated with mortal man, but only for the spiritual man. It all changed the dynamics when man sinned against God. Man was given a time of 6999 years to occupy planet earth. The spiritual beings and the spiritual worlds do not work on time limit in their dimension as the mortal man.

GOING ADRIFT:

The Gregorian calendar is continuing drifting away from nature and humanity on a yearly basis. It will always be encroaching on either the winter or the summer seasons by creating false data to the climatologists and meteorologists. A mission statement by the mainstream scientific community should be focusing on a change of calendar, but not in exhausting unnecessary monies toward a non- existing matter on climate change, which is normally the earth's progression. The added season known as autumn is the time when crops are harvested and the ant gathers its food to be stored in the harvest. Psalm 6:6-8. Spring was also added by mankind as an intermediary, prior to the summer months.

INCONSISTENCY WITH THE LUNAR CALENDAR:

The lunar calendars of those nations that use the moon as their calendar have a different amount of days calculated for their entire year. Some nation's calendar carries 360 days, while another has 354 days, even though there is only one moon revolving around planet earth. Prior to 1996 and after, the lunar calendar sometimes carries two moons in certain months, but it still observed twelve months in the year according to their religious observances.

Below are the following examples:

1996, July 1st/30th
1999, January 2nd/31st   (3 years)
2001, November 1st/30th   (2 years)
2004, July 2nd/31st   (3 years)

2007, Jun 1st/30th	(3 years)
2009, December 2nd/31st	(2 years)
2012, August 2nd/31st	(3 years)

WORLD POPULATION INCREASE

The world's population is rapidly increasing while food products are decreasing. This is due to the food consumption of an over populated world of approximately eight billion people, including unregistered illegal aliens. As a matter of fact, a very high percentage of food items on the shelves in the super markets or grocery stores can be genetic modified organisms [GMO]. Some people are resorting in purchasing organic food that is more expensive to the pocket book and wallets, but the effort is worthwhile in staying healthy. Plant seeds are known to be imported into the United States of America from India and maybe other countries. The seeds are cross pollinated making them impossible to be replanted, because its fruit is bearing without the natural seeds.

The axe is grinding, trust is out of the window and selfishness has replaced trust. This is what is considered as the shortening of days and the tribulation period before the coming of the Messiah. Below are examples of the weather inconsistencies in summarizing a claim by the scientists that climate change and global warming on planet earth are on the increase.

WEATHER HISTORY FOR JFK AIRPORT, NEW YORK, USA ON THE 20TH DAY OF EACH MONTH

CHART:

Year	Jan	Feb	Mar	Apr	May	Jun	Jul	Aug	Sep	Oct	Nov	Dec
2010	44°	45°	70°	67°	72°	91°	89°	92°	74°	60°	56°	34°
2009	29°	35°	46°	54°	73°	72°	82°	87°	72°	62°	61°	53°
2008	35°	34°	54°	63°	55°	77°	88°	75°	69°	62°	40°	32°
2007	34°	46°	52°	70°	76°	82°	82°	74°	78°	70°	49°	46°
2006	56°	36°	42°	78°	69°	83°	84°	91°	72°	66°	45°	44°
2005	31°	38°	42°	80°	55°	73°	91°	83°	82°	62°	56°	35°
2004	28°	40°	45°	76°	69°	74°	85°	85°	72°	57°	54°	21°
2003	33°	44°	48°	60°	74°	72°	82°	87°	80°	58°	52°	40°
2002	37°	50°	45°	69°	57°	77°	82°	82°	75°	59°	52°	54°
2001	37°	50°	51°	54°	59°	82°	80°	84°	73°	69°	55°	46°
1999	46°	39°	50°	48°	75°	75°	82°	77°	73°	55°	61°	57°

Year	Jan	Feb	Mar	Apr	May	Jun	Jul	Aug	Sep	Oct	Nov	Dec
1998	39°	48°	46°	60°	71°	86°	89°	78°	78°	66°	57°	50°
1997	33°	51°	46°	57°	72°	84°	77°	71°	84°	66°	53°	50°
1996	33°	48°	45°	57°	95°	70°	78°	78°	78°	59°	45°	30°
1995	50°	50°	51°	69°	72°	93°	88°	84°	73°	72°	48°	28°
1994	14°	46°	50°	64°	55°	81°	86°	80°	73°	64°	51°	45°
1993	45°	32°	36°	57°	63°	78°	90°	79°	63°	57°	53°	44°
1992	28°	51°	39°	52°	63°	73°	81°	79°	72°	48°	46°	48°
1991	55°	55°	57°	48°	64°	87°	93°	75°	66°	55°	62°	39°
1990	37°	36°	52°	55°	60°	75°	91°	66°	72°	60°	53°	45°

The inconsistency of the weather will be continuing, because calendars are lacking many days in the year. Every rotation shall be either a high or low temperature at the beginning and at the end of the year as shown on the chart. Meteorologist and climatologist would be able to review the weather commentaries; produce an accurate assessment and a comprehensive analysis with their findings.

Same: Jan 2002 - 31°; Feb 2002 - 50°; Mar 2006 - 42°; Apr 1997 - 57°
Jan 2001 - 31°; Feb 2001 - 50°; Mar 2005 - 42°; Apr 1996 - 57°

Jan 1997 - 33° _____   Mar 1998 - 46° _____
Jan 1996 - 33° _____   Mar 1997 - 46° _____

May 1993 - 63°; Jul 2003 - 82°; Aug 1993 - 79°; Sep 1995 - 73°
May 1992 - 63°; Jul 2002 - 82°; Aug 1992 - 79°; Sep 1994 - 73°

Oct 2009 - 62°; Nov 2003 - 52°; Dec 1998 - 50°
Oct 2008 - 62°; Nov 2002 - 52°; Dec 1997 - 50°

Oct 1998 - 66° _____
Oct 1997 - 66° _____

FROM 1990-2010, JAN THRU DEC, ON THE TWENTIETH DAY OF EACH MONTH

DEGREE DIFFERENCES (+ AND -)

Year	Jan	Feb	Mar	Apr	May	Jun	Jul	Aug	Sep	Oct	Nov	Dec	Comments
2010		+1°	+25°	3°	+5°	+19°	2°	+3°	18°	14°	4°	22°	
2009		+6°	+11°	+8°	+19°	1°	+10°	+5°	15°	10°	1°	8°	
2008		1°	+20°	9°	8°	+22°	+11°	13°	6°	7°	22°	8°	
2007		+12°	+6°	+18°	+6°	+6°	same	8°	+4°	8°	21°	3°	June and July 82°, The same
2006		20°	+6°	+36°	9°	+14°	+1°	+7°	19°	6°	21°	1°	
2005		7°	+4°	+38°	25°	+18°	+18°	8°	1°	20°	6°	21°	
2004		+12°	+5°	+31°	7°	+5°	+11°	same	13°	15°	3°	33°	July and August 85° The same
2003		11°	+4°	+12°	+14°	2°	+10°	+5°	7°	22°	6°	12°	
2002		13°	5°	+24°	12°	+20°	+5°	same	7°	16°	7°	2°	July and August 82° The same
2001		13°	+1°	+3°	+5°	+23°	2°	+4°	11°	4°	14°	9°	
2000													No data Recorded
1999		7°	+11°	2°	+27°	same	+7°	5°	4°	18°	+6°	4°	May and June 75°, The same
1998		9°	2°	+14°	+11°	+15°	+3°	11°	same	18°	9°	7°	August and September 78°, The same
1997		18°	5°	+11°	+15°	+12°	7°	6°	+13°	18°	13°	3°	
1996		15°	3°	+12°	+38°	25°	+8°	same	same	19°	14°	15°	July, August, and September 78°, The same
1995		same	+1°	+18°	+3°	+21°	5°	4°	11°	1°	24°	20°	January and February 50°, The same
1994		+32°	+4°	+14°	9°	+26°	+5°	6°	7°	9°	13°	6°	

Year	Jan	Feb	Mar	Apr	May	Jun	Jul	Aug	Sep	Oct	Nov	Dec	Comments
1993		13°	+4°	+21°	+6°	+15°	+12°	11°	16°	6°	4°	9°	
1992		+23°	12°	+12°	+11°	+10°	+8°	2°	7°	24°	2°	+2°	
1991		same	+2°	9°	+16°	+23°	+6°	18°	9°	11	+7°	23°	January and February 55°, The same
1990		1°	+16°	+3°	+5°	+15°	+16°	25°	+6°	12°	7°	8°	

THE TIMEKEEPER OF EVERYTHING

The Almighty God is the timekeeper of all things that He has created in the seven worlds. God has designated a certain amount of years for every living organism on planet earth. He can also add or subtract one's years to live in this world. In Isaiah 38: 5, God did add fifteen more years to King Hezekiah's year when he was gravely ill.

WINTER AND SUMMER

The two seasons that God assigned to the border of planet earth are winter and summer. Both seasons are on a timer, totaling 480 days or 480 degrees. The timer is divided up in half with the winter at 240 days and the summer at 240 days. At the beginning of the winter month in the first day of the year of 480 days, the earth begins tilting to the left at five degrees and it continues until it reaches a maximum of thirty degrees until the end of the sixth month. After the winter months are over, the earth reverses back to its position, in the beginning of the summer in the seventh month.

However, the Almighty God is able to alter the amount of degrees as He wishes to do so, as in the case of King Ahaz sun dial clock. God brought down the shadow of ten degrees on the sun dial clock, and then He reversed it back to its original position of ten degrees, Isaiah 38: 8

Man cannot complicate, confuse or change the seasons of the plants, animals, insects and other living organisms that God created on earth. Man's behavior is very hubristic, and he will not acknowledge that there is a God that created everything in the universe, and on the earth, but man does everything contrary to the word of God. The definitions of summer and winter shall not be altered, added to, change its meaning or sequence to adjust to man made calendars. The Prophetic calendar solely functions in conjunction with the seasons and the Cosmos. *Will snow actually fall in the summer? The answer is yes!* Proverbs 26:1; 25:13 states that, "as snow in summer and as rain in harvest, so honor is not seemly for a fool".

The extreme weather and climate broadcasted on the mainstream media would be inaccurate, by using the Gregorian calendar as measurements, and for time line events. God gave all living creatures or man and mankind ample time in preparing and going into another season, instead of an abrupt change in climate conditions. A fluctuation in the weather is normal in its season, and it is not considered as climate change or global warming.

GREGORIAN/ PROPHETIC CALENDAR

First Month 2010   NA- Normal Average

Chic. NA 39°f NYC NA 42°F **March 2010** 1	London NA 50° Greece NA 61° 2	Paris NA 54° 3	Israel rainfall, Nov-Mar 4	Bangladesh, Mar-Jun-Oct cool rainy monsoon 5	Lebanon rainfall Nov-Mar 6	7
Thailand year round 20°c - 27°c 8	Saudi Arabia. Monsoon Oct-Mar 9	India monsoon Oct- Mar 10	Singapore rainy season Dec- Mar 11	Yemen Mar-Aug, dust wind sandstorms 12	13	14
15	16	17	18	19	20	21
22	23	24	25	26	27	28
29	30	31	Chicago NA 49° NYC NA 52° **Apr 1** 32	Paris NA 61° 2 33	Greece NA 68° 3 34	India Apr-Oct summer 45 °+ 4 35
	NYC max 91° record previous 1947-89° 5 36	6 37	Cairo - 64° London - 55° NYC - 76° 7 38	Cairo 83° NYC 62° 8 39	9 40	

GREGORIAN/ PROPHETIC CALENDAR

Second Month 2010   NA- Normal Average

London NA 55 **April 10**th 1	11 2	Fiji Island Winter April-Sep 12 3	Switzerland Rainy April-Nov 13 4	14 5	15 6	16 7
17 8	18 9	10	20 11	21 12	22 13	Tel – Aviv 75 Rainy London 23 14
Tel – Aviv 79 Cloudy 24 15	25 16	26 17	27 18	28 19	Pakistan Spring ends 29 20	Tel – Aviv 68 Dubai 97 30 21
Pakistan Summer May - Sep **May 1**st 22	Paris NA 68 NYC 81 2 23	Denmark May - Aug 20 c+ 3 24	4 25	Ireland May - Sep; dry; 15c – 25c 5 26	Switzerland May, Sep and part of Oct dry; 30c 6 27	7 28
London NA 63 Greece NA 77 8 29	9 30	10 31	11 32	12 33	13 34	14 35
15 36	16 37	17 38	18 39	19 40		

GREGORIAN/ PROPHETIC CALENDAR

<u>Third Month 2010</u>   NA- Normal Average

Vietnam south May-Oct, hot 20 1	21 2	23 3	24 4	25 5	26 6	27 7
  28 8	29 9	30 10	31 11	32 12	**June 1** 13	London NA 66° Paris NA 73° 2 14
Greece NA 86°f 3 15	Israel rainless Jun-Aug 4 16	Cambodia, Jun-Oct rainy season 5 17	NYC 85° 6 18	Bangladesh, Oct-Mar dry winter cool 7 19	Belarus, summer month 25° c- 30°c 8 20	Iran, Jun-Aug no rain 9 21
Lebanon, no rain Jun-Aug 10 22	11 23	New Zealand, winter in June 7°c 12 24	Yemen Summer Jun- Sep 40°C+ 13 25	Yemen, Jun-Sep rain fall 14 26	Singapore, dry season Jun- Sep 15 27	16 28
  17 29	18 30	19 31	20 32	21 33	22 34	23 35
  24 36	25 37	26 38	27 39	28 40		

GREGORIAN/ PROPHETIC CALENDAR

Fourth Month 2010   NA- Normal Average

			London NA 70° Paris NA 77°	Greece NA 91° Jul-Aug 30°c- 35°ct	Australia Jul–Aug coldest months	England & Scotland Jul- Aug 16°c - 21°c
29 1	30 2	**July 1**	2 4	3 5	4 6	5 7
Finland Jul 13°c - 17°c	Germany warmest month 20°ct	Iceland Jul 11.2°c	New Zealand Jul coldest	Norway Jul average 16°c	Russia hot & dry in summer	Saudi Arabia summer 45°c- 54°c
6 8	7 9	8 10	9 11	10 12	11 13	12 14
Syria summer 31°c	Turkey summer average 17°c-23°c					
13 15	14 16	15 17	16 18	17 19	18 20	19 21
20 22	21 23	22 24	23 25	24 26	25 27	26 28
						London NA 72° Paris NA 75°
27 29	28 30	29 31	30 32	31 33	**Aug 1** 34	2 35
Greece NA 91°	Israel no rain Jun-Aug 18°c - 38°c	Iran summer 20° - 30°ct	Iraq Aug – hottest 30° - 45°c	Italy summer 28° - 40°c		
3 36	4 37	5 38	6 39	7 40		

GREGORIAN/ PROPHETIC CALENDAR

Fifth Month 2010   NA- Normal Average

Lebanon Aug- Hottest 18°c - 40°c 8 — 1	Lebanon no rain Jun- Aug 9 — 2	10 — 3	11 — 4	12 — 5	13 — 6	14 — 7
15 — 8	16 — 9	17 — 10	18 — 11	19 — 12	20 — 13	21 — 14
22 — 15	23 — 16	24 — 17	25 — 18	26 — 19	27 — 20	Tel- Aviv 68° Dubai 97° 28 — 21
29 — 22	30 — 23	31 — 24	**Sep 1** — 25	London NA 66° Paris NA 70° 2 — 26	Greece NA 84° 3 — 27	Denmark, rainy in Sep 4 — 28
5 — 29	Yemen Jun- Sep, rainfall 6 — 30	7 — 31	8 — 32	9 — 33	New moon and Sabbath; year of 5507 10 — 34	11 — 35
12 — 36	13 — 37	14 — 38	15 — 39	16 — 40		

GREGORIAN/ PROPHETIC CALENDAR

<u>Sixth Month 2010</u>   <u>NA- Normal Average</u>

			London 21°c	London 21°c	NYC 85° London 22°c Israel 85° Iran 86°	NYC 82° London 19°c Iran 90° Russia 11°c
17 1	18 2	19 3	20 4	21 5	22 6	23 7
NYC 88° London 14°c Israel 87°	NYC 75° Iran 84°	NYC 67° Iran 86°	NYC 72° London 15°c Russia 18°c	NYC 73° London 16°c Israel 81°	NYC 72° London 15°c	Pakistan, end of summer
24 8	25 9	26 10	27 11	28 12	29 13	30 14
NYC 69° London 15°c Israel 84°	London NA 57° Paris NA 61°	Greece NA 75°		Saudi Arabia, India monsoon Oct-Mar		Russia's snow starts in Oct month
Oct 1 15	2 16	3 17	4 18	5 19	6 20	7 21
8 22	9 23	10 24	11 25	12 26	13 27	14 28
15 29	16 30	17 31	18 32	19 33	20 34	21 35
			NYC 71°F	NYC 75°F		
22 36	23 37	24 38	25 39	26 40		

GREGORIAN/ PROPHETIC CALENDAR

Seventh Month 2010   NA- Normal Average

NYC 73°F 27 1	NYC 74°F 28 2	 29 3	 30 4	 31 5	49°F **Nov 1** 6	London NA 50°F; Paris NA 50°F NYC 49°F 2 7
Greece NA 66°F Israel, rainfall Nov- Mar 3 8	Fiji Isl Nov-Feb Summer, less than 35°c 4 9	Lebanon rainfall Nov- Mar 5 10	Turkey winter Nov-Apr 13°c 6 11	 7 12	 8 13	 9 14
  10 15	 11 16	 12 17	 13 18	 14 19	 15 20	 16 21
  17 22	 18 23	 19 24	 20 25	 21 26	 22 27	 23 28
  24 29	 25 30	 26 31	 27 32	 28 33	 29 34	 30 35
Pakistan winter Dec- Feb **Dec 1** 36	London NA 45°F Paris NA 45°F 2 37	Greece NA 59°F 3 38	Australia, Dec- Jan hottest months 4 39	Cambodia, Dec- Mar, dry season 5 40		

GREGORIAN/ PROPHETIC CALENDAR

Eight Month 2010-2011   NA- Normal Average

India, Dec-Jan 5°C		Singapore rainy season Dec- Mar 28°c	New Zealand summer starts in Dec			
6 1	7 2	8 3	9 4	10 5	11 6	12 7
						(2010) NYC 20" Blizzard
13 8	14 9	15 10	16 11	17 12	18 13	19 14
20" Blizzard NYC	London NA 45°F Paris NA 43°F				Jerusalem snow Jan- Feb 5°c- 10°c 42° New year	
20 15	21 16	22 17	23 18	24 19	25 20	26 21
Iraq coldest in Jan 5°c- 10°c	Greece NA 55°F				Iceland Jan average temp 0.4°c	
27 22	28 23	29 24	30 25	Dec 31 2010 26	Jan 1 2011 27	2 28
Lebanon coldest Jan 5°c- 10°c	New Zealand Jan- Feb warmest 16°c - 30°c	Norway average 10°c	Russia Jan- 50° c	Syria winter 4°c		
3 29	4 30	5 31	6 32	7 33	8 34	9 35
10 36	11 37	12 38	13 39	14 40		

GREGORIAN/ PROPHETIC CALENDAR

<u>Ninth Month 2011</u>   <u>NA- Normal Average</u>

	Lebanon no rain Jun- Aug					
15 1	16 2	17 3	18 4	19 5	20 6	21 7
22 8	23 9	24 10	25 11	26 12	27 13	28 14
				London NA 46°F Paris NA 45°F		
29 15	30 16	31 17	**2011 Feb 1** 18	2 19	3 20	4 21
Finland coldest 22°c- 50°c						
5 22	6 23	7 24	8 25	9 26	10 27	11 28
12 29	13 30	14 31	15 32	16 33	17 34	18 35
19 36	20 37	21 38	22 39	23 40		

GREGORIAN/ PROPHETIC CALENDAR

Tenth Month 2011   NA- Normal Average

				Pakistan winter ends		
24 · 1	25 · 2	26 · 3	27 · 4	28 · 5	**Mar 1** · 6	2 · 7
3 · 8	4 · 9	5 · 10	6 · 11	7 · 12	8 · 13	9 · 14
10 · 15	11 · 16	12 · 17	13 · 18	14 · 19	15 · 20	16 · 21
17 · 22	18 · 23	19 · 24	20 · 25	21 · 26	22 · 27	23 · 28
24 · 29	25 · 30	26 · 31	27 · 32	28 · 33	29 · 34	30 · 35
31 · 36	**Apr 1 2011** · 37	2 · 38	3 · 39	4 · 40		

Gregorian/ Prophetic Calendar

Eleventh Month 2011   NA- Normal Average

5 1	6 2	7 3	8 4	9 5	10 6	11 7
12 8	13 9	14 10	15 11	16 12	17 13	18 14
19 15	20 16	21 17	22 18	23 19	24 20	25 21
26 22	27 23	28 24	29 25	30 26	May 1 27	2 28
3 29	4 30	5 31	6 32	7 33	8 34	9 35
10 36	11 37	12 38	13 39	14 40		

GREGORIAN/ PROPHETIC CALENDAR

Twelfth Month 2011   **NA- Normal Average**

15 — 1	16 — 2	17 — 3	18 — 4	19 — 5	20 — 6	21 — 7
22 — 8	23 — 9	24 — 10	25 — 11	26 — 12	27 — 13	28 — 14
29 — 15	30 — 16	31 — 17	Jun 1 2011 — 18	2 — 19	3 — 20	4 — 21
5 — 22	6 — 23	7 — 24	8 — 25	9 — 26	10 — 27	11 — 28
12 — 29	13 — 30	14 — 31	15 — 32	16 — 33	17 — 34	18 — 35
19 — 36	20 — 37	21 — 38	22 — 39	23 — 40		

Note: The tenth; eleventh and twelfth months on the Prophetic calendar do not have weather data, because the Gregorian calendar is shorting thirty days on the Prophetic calendar, and ends in the ninth month on the Prophetic calendar.

CHAPTER FIVE

The Prophetic moon matrix 21 years cycle.   SM - Single Moon; DM – Double Moon

Year	1	2	3	4	5	6	7	8	9	10	11	12	Moon	Year
1	4/34 Start	23	12	2/31	20	10/40	30	20	10/40	30	19	9/39	17	5489
2	28	18	7/37	26	16	6/35	25	15	5/34	23	13	2/32	16	5490
3	21	11/40	30	20	10/39	29	18	7/37	26	14	2/32	20	16	5491
4	10/40	28	19	8/38	27	17	6/36	25	15	4/34	23	12	16	5492
5	27	17	6/36	25	15	4/34	23	12	2/31	21	11	1/30	16	5493
6	19	9/38	28	17	6/36	25	15	4/34	24	13	2/31	21	16	5494
7	10/39	29	18	8/37	27	17	7/36	26	15	5/34	23	13	16	5495
8	2/31	21	11	1/30	20	10/39	29	18	7/37	26	15	5/35	17	5496
9	24	14	4/34	23	13	2/31	21	10/39	29	19	8/37	27	16	5497
10	16	6/35	25	14	4/33	23	13	3/32	22	11	1/31	20	16	5498
11	10/39	29	18	8/37	27	16	5/34	24	13	3/33	22	12	16	5499
12	1/31	20	9/38	28	17	7/37	26	16	6/36	25	15	4/34	17	5500
13	23	13	2/32	21	11	1/30	20	10/39	29	18	7/37	26	16	5501
14	16	6/36	25	15	4/34	23	13	2/32	21	11/40	30	19	16	5502
15	9/39	28	18	7/37	26	16	6/35	25	15	5/34	23	13	16	5503
16	2/32	21	11/40	30	20	10/39	29	18	7/37	26	14	3/32	17	5504
17	20	10/40	28	19	8/38	27	17	6/36	25	15	4/34	23	16	5505
18	12	27	17	6/36	25	15	4/34	23	12	2/31	21	11/40	16	5506
19	30	19	9/38	28	17	6/36	25	15	4/34	24	13	2/31	16	5507
20	21	10/39	29	18	8/37	27	17	7/36	26	15	5/34	23	16	5508
21	13	2/31	21	11	1/30	20	10/39	29	18	7/37	26	15* End	16	5509
End	29	28	28	28	28	30	27	29	28	28	28	29	340M	21Yr Cycle

"Where there is an end, there's a new beginning."

In adding the vertical and lateral numbers on the full moon chart they both carry the same amount of 340 full moons in the twenty one year Prophetic cycle.

In the first month on the Prophetic calendar in the 21 year moon cycle, Yehshua was crucified on the 14th day in the year 3979. This event took place on that actual date as it is indicated on the chart (13th). In comparing the date on the Gregorian calendar, it is October 30th 2012. The next flood - storm that shall affect the Tri-State areas in the USA will be in January 4th 2038- 26 years and 3 months after October 30th 2012. 20 years × 480 days = 9,600 days; 9,600 ÷ 365 = 26.3 years

A new moon cycle is expected to be ending in December 17th, 2013 / 5509. A twenty one year- cycle on the Prophetic calendar is twenty seven years and six months on the Gregorian calendar. So then, the ending of the next cycle shall be June 16th 2040. (Prophetic 21years × 480 days = 10080 days; 10080 ÷ 365 = 27.616838 years in accordance with the Gregorian time-line

ORBITAL TRAITS

Hypothesis: The Cosmos 21 year cycle incorporates both sun and the moon's motions. The sun's gyration around the universe or heaven is 480 days, and the moon's revolution around the Cosmos in a 21 year cycle is a total of 340 full moons. Even though, the moon is in its thick darkness in the Cosmos and without its shine - it is still in motion, but still has an effect on planet earth. So then, a 21 year cycle of the sun and the moon is:

The sun's 480 day × 21 years = 10,080 days

The 340 full moons × 21 year = 7,140 days

Total = 17,220 days

A SOLAR MONTH

In actuality, a solar month according to astronomers consists of 28 days, and it has 13 months in a year - even though the sun should not be used as a calendar. If the scientific community is convinced that a solar year has twelve months, so then, one year should be 28 days × 12 months = 336 days – not 365 days.

In a 21 year cycle of 336 days = 7,056 days

In a 21 year full moons = 7,140 days

Total = 14,196 days

At an average assessment of a full moon appearance on planet earth at 14,196 ÷ 480 = 29.575 days according to the scientific mathematical calculation.

COUNTDOWN TO THE END OF THE AGE

Prophetic calendar Year	Gregorian calendar Year.
550 9/15/12	12/ 14/2013
5530/15/12	June 2040
5551/15/12	December 2067
5572/15/12	June 2094
5593/15/12	December 2121
5614/15/12	June 2148
5635/15/12	December 2175
5656/15/12	June 2202
5677/15/12	December 2229
5699/15/12	June 2256
5720/15/12	December 2286
5741/15/12	June 2316
5762/15/12	December 2343
5783/15/12	June 2370
5804/15/12	December 2397
5825/15/12	June 2418
5846/15/12	December 2445
5867/15/12	June 2472
5888/15/12	December 2499
5909/15/12	June 2526
5930/15/12	December 2553
5951/15/12	June 2580
5972/15/12	December 2607
5992/14/01	* 2633 Satan imprisoned
5993/15/12	June 2634
6014/15/12	December 2661
6035/15/12	June 2688
6056/15/12	December 2715
6077/15/12	June 2742
6098/15/12	December 2769
6119/15/12	June 2796
6140/15/12	December 2823
6161/15/12	June 2850

6182/15/12	December 2877
6203/15/12	June 2904
6224/15/12	December 2931
6245/15/12	June 2958
6266/15/12	December 2985
6287/15/12	June 3012
6308/15/12	December 3039
6329/15/12	June 3066
6350/15/12	December 3093
6371/15/12	June 3129
6392/15/12	December 3147
6423/15/12	June 3174
6444/15/12	December 3201
6465/15/12	June 3228
6486/15/12	December 3255
6507/15/12	June 3282
6528/15/12	December 3309
6549/15/12	June 3336
6570/15/12	December 3363
6591/15/12	June 3390
6612/15/12	December 3417
6633/15/12	June 3444
6654/15/12	December 3471
6675/15/12	June 3498
6696/15/12	December 3525
6717/15/12	June 3552
6738/15/12	December 3579
6759/15/12	June 3806
6780/15/12	December 3833
6801/15/12	June 3860
6822/15/12	December 3887
6843/15/12	June 3914
6864/15/12	December 3941
6885/15/12	June 3969
6906/15/12	December 3996
6927/15/12	June 4023
6948/15/12	December 4050

6969/15/12   June 4077

6990/15/12   December 4104

6992/14/01   Satan released

In the year 6993 shall be the last revolving moon around planet earth. It will be the 333rd full moon before a 21 year cycle ending at 340 full moons. On the Gregorian calendar it will be in the year 4107. However, there are certain disasters destined for planet earth, which cannot be prevented. The only way to stop such a thing from happening is, that humankind has to change their destructive ways to the planet, and ask forgiveness from the Almighty God of all worlds. In Jonah 3:9, 10, the entire city and leaders did make a supplication to the Almighty God to redeem them, and they were saved from total destruction of their city.

THE GREGORIAN CALENDAR'S 19 YEARS FULL MOON MATRIX

YEAR	JAN	FEB	MAR	APR	MAY	JUN	JUL	AUG	SEP	OCT	NOV	DEC
2007	3	2	3	2	2	1/30	30	28	26	26	24	24
2026	3	2	3	2	2	1/30	30	28	26	26	24	24
2008	22	21	21	20	20	18	18	16	15	14	13	12
2027	22	21	21	20	20	18	18	16	15	14	13	12
2009	11	9	11	9	9	7	7	6	4	4	2	2/31
2028	11	9	11	9	9	7	7	6	4	4	2	2/31
2010	30	28	30	28	27	26	26	24	23	23	21	21
2029	30	28	30	28	27	26	26	24	23	23	21	21
2011	19	18	19	18	17	15	15	13	12	12	10	10
2030	19	18	19	18	17	15	15	13	12	12	10	10
2012	9	7	8	6	6	4	3	2/31	30	29	28	28
2031	9	7	8	6	6	4	3	2/31	30	29	28	28
2013	27	25	27	25	25	23	22	21	19	18	17	17
2032	27	25	27	25	25	23	22	21	19	18	17	17
2014	16	15	17	15	14	13	12	10	9	8	7	7
2033	15	14	16	14	14	13	12	10	9	8	6	6
2015	5	4	5	4	3	1	1/30	28	27	26	25	24
2034	4	3	5	3	3	2	1/31	29	28	27	25	25
2016	23	22	24	22	22	20	20	18	16	16	14	14
2035	23	22	23	22	22	20	20	19	17	17	15	15
2017	12	11	13	11	11	10	9	8	6	5	4	3
2036	13	11	12	10	10	8	8	7	5	5	4	3

YEAR	JAN	FEB	MAR	APR	MAY	JUN	JUL	AUG	SEP	OCT	NOV	DEC
2018	2/31		2/31	30	30	28	28	26	25	24	23	22
2037	2/31		2/31	29	29	27	27	25	24	24	22	22
2019	21	19	20	18	18	16	16	15	13	13	11	11
2038	21	19	21	19	18	17	16	14	13	13	11	11
2020	9	8	9	8	7	6	5	4	2	2	1/30	30
2039	10	9	10	9	8	6	6	4	2	2/31	30	30
2021	28	27	28	27	26	24	24	22	21	21	20	19
2040	29	28	28	27	26	24	24	22	20	20	18	18
2022	18	16	18	16	16	14	13	12	10	10	9	8
2041	17	16	17	16	16	14	13	12	10	9	8	7
2023	7	6	6	5	4	3	2/31	30	28	28	26	26
2042	6	5	6	5	5	3	3	1/31	29	28	27	26
2024	25	24	25	24	23	22	21	19	18	17	16	15
2043	25	23	25	24	24	22	22	20	19	18	16	16
2025	14	13	14	13	13	11	11	9	7	7	5	5
2044	14	13	13	12	12	10	10	8	7	7	5	5
2026	3	2	3	2	2	1/30	30	28	26	26	24	24
2045	3	1	3	1	1/30	29	28	27	26	25	24	24

THE MOON PHASES

The first, but only calendar introduced into this world was the Prophetic calendar. The lunar calendar was a replacement to the Prophetic calendar, when Canaan had died in the Prophetic year two hundred and seventy (lunar year 360). The Julian calendar was then crafted in 45BC or 46BC, but subsequently it became inaccurate; and it was replaced with the Gregorian calendar in 1582 – to date it is still in use.

The appearances of the full moons from 2007 to 2032 in a nineteen year cycle are parallel, but it became very inconsistent from the year2014 to 2045. There is no continuity or consistency in the Gregorian calendar nineteen year cycle. The Prophetic calendar works well in conjunction with the moon cycle, the sun and the other worlds. It has a twenty one year cycle versus the Gregorian calendar, that has a nineteen year cycle which becomes inaccurate after a certain amount of years – the cycle has no connecting dates after nineteen years, but it is still considered a cycle. The Prophetic calendar, in comparison to the Gregorian calendar has a consistent cycle as follows: On the 29th day of the first month throughout the twenty one year cycle, the full moon will appear; on the 28th of the second month, third, fourth and the fifth; the 30th day in the sixth month; 27th in the seventh month; 29th in the eight month and back to the

28th in the ninth, tenth and the eleventh months. However, in the twelfth month the full moon appears on the 29th of the month.

THE REAL MOON CYCLE

God had made two luminaries on the fourth day of creation. They were not made for the purpose of creating an almanac or to facilitate calendar days, months or a year. The heavenly luminaries came into existence to give light and to function in other capacities on planet earth. **Deuteronomy 17:3** states, that Yahweh did not command that the sun or moon is to be worshipped or served in any other ways. **Deuteronomy 4:19 concurred that,** unless you raise your eyes to heaven and you see the sun, moon, stars and the entire hosts of heaven, you may be tempted to worship them or serve them, which the Creator has distributed unto all nations under the heavens. The prophets said that the sun, stars and moon shall fall to earth at the end of days. In **Psalm 104:19** it states, that the moon is created for seasons and not for months in the year. God creations are coming to an end in the year 6997 when judgment shall be to all living souls along with the judgment of the Satan.

Number	Cycle Years
1	21
2	42
3	63
4	84
5	105
6	126
7	147
8	168
9	189
10	210
11	231
12	252
13	273
14	294
15	315

After the death Cain, the only son of Lucifer in the year 270 according to the Prophetic calendar, the lunar calendar with 360 days was adapted by the Canaanite nation. The lunar calendar was subsequently handed down to the Babylonians; and it was adapted by many other nations to date. Many religious experts along with the scientific community are claiming that the moon has a nineteen year cycle, but has to be adjusted by adding a leap month, because God's festivals were slowly drifting away year after year. This was creating a problem of inconsistency on an annual basis; and the festivals of the Creator had to be falling on different dated from time to time.

Now with the new covenant in place, God presently does not acknowledge what was done in the old covenant which is considered obsolete on planet earth.

In the year 2013 / 5509, at the end of a twenty one year moon cycle, it will be a total of 340 moons. The moon and the sun shall then be going through its phases accordingly, in the evening on its fourth day of creation as per the Prophetic calendar.

RECREATION OF THE EARTH:

The age of a recreated earth to date is 5509/10 years old, (2013). The renewal of the earth shall be with fire in the year 6999 and its new creation shall be in the year 7000. The remaining years to the end of the age or cycle is 6999 -5509=1,490 years, unless it is shortened by the Creator. Be aware that the Satan shall be taken down in chain to the bottomless pit for 1000 years in the year 5992; or 2633 on the Gregorian calendar. Therefore the ages of the seven heavens according to the Jubilee calendar are as follow:

Earth's sabbatical years are 49 years and the Jubilee years are 50 years. Heavenly sabbatical years are 49 thousand, and its Jubilee years are thousand - the same concept as the earth sabbatical and Jubilee year, except that God's time is 1000 years ahead of planet earth's time. So then, by working in conjunction with the earth's Jubilee years, there are 1,490,000 years remaining, before the last the end of the age and a Jubilee in 7,000,000 years.

The earthly years are 5510 + 1,490 remaining years to its Jubilee in 7,000.year. Isaiah 48: 13 states, that the right hand of God had made the foundation of the earth, and that His right hand had spanned out the heavens. It is indicating that the other heavens are spanned out at 8,640 hours or at 360 degrees north. Psalm 102:25; Nehemiah 9:6; Hebrew 1:10; Psalm 148:4; Isaiah 13:13; Matthew 24:29.

The 15th of the twelfth month 5509 is the last day on a twenty one year cycle. The Prophetic day is always begins at 6:00 pm and not midnight as the Gregorian days.

2 Peter 3:8; Psalm 90:4 states that one day with the Lord as a thousand years, and a thousand years as one day. To date, all seven heavens (worlds) in this cycle are 47,510,000 years old. Earth's age in its cycle is 47,510 years.

MOTION OF THE MOON

The moon orbits around the earth in its circuit on the following timetable during its twenty one year cycle:

- In the first, eight and twelfth months the moon is full on the twenty ninth day.
- In the sixth month the moon is full on the thirtieth day.
- In the seventh month the moon is full on the twenty seventh day.-In the second; third; fourth; fifth; ninth; tenth and the eleventh months, the moon is full on the twenty eight day.
- Every first, eight, twelfth and sixteenth year of the twenty one year cycle, there are a total of seventeen moons in each of the years.
- There are a total of three hundred and forty full moons in the cycle.
- The sun and the moon are working in conjunction in their separate circuits around the cosmos respectively.
- The moon is considered barren, which never produced any living organisms, but functions as a stabilizer to the earth along with the sun. It also controls the seasons on planet earth.
- The sun and moon is a disc- like shape of the same size, but the sun is seven times brighter than the moon.

The earth is the only planet that produces life and water. The other six worlds are spiritual worlds. Their inhabitants are able to do space travel to other planets including earth, because they are spiritual beings that can transform themselves into matter, and retransform back to their normal being, and they do not have a soul.

The earth's rotating speed in covering the northern and the southern hemispheres are 960 mph. The seven heavens in motion are moving at 960,000 mph into the vast empty space without an end, but only in total darkness.

The heavens move along every 1000 years at 11,520,000 hours. God said, "I will shake the heavens and the earth shall remove out of place." That will be the only time when the earth will be moving off its axis.

The circumference of the seven heavens is 161,280,000 approximate miles. (23,040,000 miles × 7). *In 2 Peter 3:10 states that the heavens shall pass away with a great noise and the elements shall melt with fervent heat.* The earth also, and the works that are within shall be burned up.

CHAPTER SIX

A Jubilee calendar, re dating, prophesies and the end of the **age.**

Note: Not all Millennium Years are aligned With Jubilee years.

480 Days	The Jubilee calendar beginning in the year four hundred and eighty. Things to be brought forth were already in place in that year awaiting God's command.	
First Month of the year 01	Creation started in the first month. **Genesis 2:1**, the Heavens and the earth were finished and all hosts of them. **Genesis 2:7** Adam was created an adult and as a matured man at age 30 years old. **Genesis 2:21** Eve was created at a matured age; out of the flesh and bone of Adam; Rest Day7.	The earth was there, but uninhabited; dark and only water on its surface. New moon, sun and other luminaries created. 1/4/01
007	The first Sabbatical year.	7+7 1
014		14+7 2
021		21+7 3
028		28+7 4
035		35+7 5
039	Eve was deceived by the serpent Lucifer in the garden of Eden.	Gen 3:6
039	"I will put enmity between your seed and her seed."	Gen 3:15
040	Cain was born from the seed of Lucifer.	Gen 4:1
041	Abel was born from Adam's seed.	Gen 4:2
042		42+7 6

049	Sabbatical year		7
050	First Jubilee year	Adam was 80 years old	one
057		57+7	1
064		64+7	2
071		71+7	3
078		78+7	4
080	Cain at forty years old		
081	Abel at thirty nine years old		
085		85+7	5
092	Cain disliked Abel; but he was instigated by his father, Lucifer to commit murder.	92+7	6
099	Sabbatical year		7
099	Cain and Abel's first time offerings to God. Cain offered up fruits, but Abel offered up the first born of his flock. Num 18:12.		
100	Cain killed his brother Abel, because of jealousy.		
100	Jubilee year	Two	
107		107+7	1
114		114+7	2
121		121+7	3
128		128+7	4
130	Seth born unto Adam	Gen 5:3	
135		135+7	5
142		142+7	6
149	Sabbatical year		7
150	Jubilee year		3
157		157+7	1
164		164+7	2
171		171+7	3
178		178+7	4
185		185+7	5
192		192+7	6
199	Sabbatical year		7
200	Jubilee year		4
207		207+7	1
214		214+7	2

221		221+7	3
228		228+7	4
235	Enos born unto Seth	Gen 5:6	
242		242+7	6
249	Sabbatical year		7
250	Jubilee year		5
257		257+7	1
264		264+7	2
270	Cain died at age 270 years old. (360 lunar years)		
271		271+7	3
278		278+7	4
285		285+7	5
292		292+7	6
299	Sabbatical year		7
300	Jubilee year		6
307		307+7	1
314		314+7	2
321		321+7	3
325	Cainan born unto Enos	Gen 5:9	
328		328+7	4
335		335+7	5
342		342+7	6
349	Sabbatical year		7
350	Jubilee year		7
357		357+7	1
364		364+7	2
371		371+7	3
378		378+7	4
385		285+7	5
392		392+7	6
395	Mahalaleel born unto cainan	Gen 5:12	
399	Sabbatical year		7
400	Jubilee year		8
407		407+7	1
414		414+7	2
421		421+7	3
428		428+7	4

435		435+7	5
442		442+7	6
449	Sabbatical year		7
450	Jubilee year		9
457		457+7	1
460	Jared born unto Mahalaleel	Gen 5:15	
464		464+7	2
471		471+7	3
478		478+7	4
485		485+7	5
492		492+7	6
499	Sabbatical year		7
500	Jubilee year. Shem died. Lunar years 666.66666		10
507		507+7	1
514		514+7	2
521		521+7	3
528		528+7	4
535		535+7	5
542		542+7	6
549	Sabbatical year		7
550	Jubilee year		11
557		557+7	1
564		564+7	2
571		571+7	3
578		578+7	4
585		585+7	5
592		592+7	6
599	Sabbatical year		7
600	Jubilee year		12
607		607+7	1
614		614+7	2
621		621+7	3
622	Enoch born unto Jared	Gen 5:18	
628		628+7	4
635		635+7	5
642		642+7	6
649	Sabbatical year		7

650	Jubilee year		13
657		657+7	1
664		664+7	2
671		671+7	3
678		678+7	4
685		685+7	5
692		692+7	6
699	Sabbatical year		7
700	Jubilee year		14
707		707+7	1
714		714+7	2
721		721+7	3
728		728+7	4
735		735+7	5
742		742+7	6
749	Sabbatical year		7
750	Jubilee year		15
757		757+7	1
764		764+7	2
771		771+7	3
778		778+7	4
785		785+7	5
792		792+7	6
799	Sabbatical year		7
800	Jubilee year		16
807		807+7	1
814		821+7	2
821		821+7	3
828		828+7	4
835		835+7	5
842		842+7	6
849	Sabbatical year		7
850	Jubilee year		17
857		857+7	1
864		864+7	2
871		871+7	3
874	Lamech born unto Methuselah	Gen 5:25	

878		878+7	4
885		885+7	5
892		892+7	6
899	Sabbatical year		7
900			18
907		907+7	1
914		914+7	2
921		921+7	3
928		928+7	4
930	Daughters and sons born unto Adam; Adam died at 930 years.	<u>Gen 5:5</u>	
935		935+7	5
937	Sons and daughters born unto Seth	<u>Gen 5:7</u>	
942		942+7	6
949	Sabbatical year		7
950	Jubilee year		19
957		957+7	1
964		964+7	2
970			
971		971+7	3
978		978+7	4
985		985+7	5
987	Sons and daughters born unto Enoch; Enoch age 365 years was taken.	<u>Gen 5:24</u> 365 years on Prophetic Calendar, and 480 years on the Gregorian Calendar	
992		992+7	6
999	Sabbatical year		7
1000	Jubilee year		20
1007		1007+7	1
1014		1014+7	2
1021		1021+7	3
1028		1028+7	4
1035		1035+7	5

1042		1042+7	6
1042	Seth died at age 912 years	Gen 5:8	
1049	Sabbatical year		7
1050	Jubilee year		21
1056	Noah born unto Lamech	Gen 5:28	
1057		1057+7	1
1064		1064+7	2
1071		1071+7	3
1078		1078+7	4
1085		1085+7	5
1092		1092+7	6
1099	Sabbatical year		7
1100	Jubilee year		22
1107		1107+7	1
1114		1114+7	2
1121		1121+7	3
1128		1128+7	4
1135		1135+7	5
1140	Sons and daughters born unto Enos. He died at age 905 years.	Gen 5:11	
1142		1142+7	6
1149	Sabbatical year		7
1150	Jubilee year		23
1157		1157+7	1
1164		1164+7	2
1171		1171+7	3
1178		1178+7	4
1185		1185+7	5
1192		1192+7	6
1199	Sabbatical year		7
1200	Jubilee year		24
1207		1207+7	1
1214		1214+7	2
1221		1221+7	3
1228		1228+7	4
1231 1235	Sons and daughters born unto Cainan; he died at age 910 years.	Gen 5:14 1235+7	5

1242		1242+7	6
1249	Sabbatical year		7
1250	Jubilee year		25
1257		1257+7	1
1264		1264+7	2
1271		1271+7	3
1278		1278+7	4
1285		1285+7	5
1290	Sons, and daughters born unto Mahalaleel; He died at age 895 years.	Gen 5:17	
1292		1292+7	6
1299	Sabbatical year		7
1300	Jubilee year		26
1307		1307+7	1
1314		1314+7	2
1321		1321+7	3
1328		1328+7	4
1335		1335+7	5
1342		1342+7	6
1349	Sabbatical year		7
1350	Jubilee year		27
1357		1357+7	1
1364		1364+7	2
1371		1371+7	3
1378		1378+7	4
1385		1385+7	5
1392		1392+7	6
1399	Sabbatical year		7
1400	Jubilee year		28
1407		1407+7	1
1414		1414+7	2
1421		1421+7	3
1422	Sons, and daughters born unto Jared; He died at age 962 years.	Gen 5:20	
1428		1428+7	4
1435		1435+7	5
1442		1442+7	6

1449	Sabbatical year		7
1450	Jubilee year		29
1457		1457+7	1
1464		1464+7	2
1471		1471+7	3
1478		1478+7	4
1485		1485+7	5
1492		1492+7	6
1499	Sabbatical year		7
1500	Jubilee year		30
1507		1507+7	1
1514		1514+7	2
1521		1221+7	3
1528		1528+7	4
1535		1535+7	5
1536	Yahweh warned the people on earth 120 years prior the flood.	Gen 6:3 480 x 120 years = 57,600 days	
1542		1542+7	6
1549	Sabbatical year		7
1550	Jubilee year		31
1556	Shem, Ham and Japheth were born unto Noah. (Triplets)	Gen 5:32 7:13 Gen 6:10	
1557		1557+7	1
1564		1564+7	2
1571		1571+7	3
1578		1578+7	4
1585		1585+7	5
1592		1592+7	6
1599	Sabbatical year		7
1600	Jubilee year		32
1607		1607+7	1
1614		1614+7	2
1621		1621+7	3
1628		1628+7	4
1635		1635+7	5
1642		1642+7	6

1649	Sabbatical year		7
1650	Jubilee year		33
1651	Sons, and daughters born unto Lamech; He died at age 777 years	Gen 5:31	
1656	Methuselah got sons and daughters; Methuselah died at 969 years; Noah was 600 years old in the year of the flood. Lunar years are 789.	Gen 5:26 Gen 5:27	
1657		1657+7	1
1658	Two years after flood, Arphaxad was born unto Shem	1Chr 1:17 Gen 11:10	
1664		1664+7	2
1671		1671+7	3
1678		1678+7	4
1685		1685+7	5
1692		1692+7	6
1693	Selah born unto Arphaxad	Gen 11:12 Luke 3:35	
1699	Sabbatical year		7
1700	Jubilee year		34
1707		1707+7	1
1714		1714+7	2
1721		1721+7	3
1723	Eber born unto Selah	Gen 11:14	
1728		1728+7	4
1735		1735+7	5
1742		1742+7	6
1749	Sabbatical year		7
1750	Jubilee year		35
1757	Peleg born unto Eber	Gen 11:16	
1757		1757+7	1
1764		1764+7	2
1771		1771+7	3
1778		1778+7	4
1785		1785+7	5
1787	Rue born unto Peleg	Gen 11:18	
1792		1792+7	6
1799	Sabbatical year		7

1800	Jubilee year		36
1807		1807+7	1
1814		1814+7	2
1819	Serug born unto Rue	Gen 11:20	
1821		1821+7	3
1828		1828+7	4
1835		1835+7	5
1842		1842+7	6
1849	Sabbatical year		7
1850	Jubilee year		37
1857		1857+7	1
1864		1864+7	2
1871		1871+7	3
1878		1878+7	4
1878	Terah born unto Nahor	Gen 11:24	
1885		1885+7	5
1892		1892+7	6
1899	Sabbatical year		7
1900	Jubilee year		38
1907		1907+7	1
1914		1914+7	2
1921		1921+7	3
1928		1928+7	4
1935		1935+7	5
1942		1942+7	6
1948	Abram, Nahor and Haran born. (Triplets)	Josh 24:2	
1949	Sabbatical year		7
1950	Jubilee year		39
1957		1957+7	1
1962	Sarah born	1962+7	2
1971		1971+7	3
1978		1978+7	4
1985		1985+7	5
1992		1992+7	6
1994	Sons and daughters born unto Rue; He died at 207 years.	Gen 11:21	

1996	Sons and daughters born unto Peleg; He died at 209 years.	Gen 11:19
1997	Sons and daughters born unto Nahor; He died at 148 years.	Gen 11:25
1999	Sabbatical year	7
2000	Jubilee year	40
2006	After the flood, Noah died at 950 years old. Lunar years are 1,266.66666	Gen 9:28-29
2007		2007+7 1
2014		2014+7 2
2019	Sons and daughters born unto Serug; He died at 200 years. Lunar years are 266.66666. Serug is from a seed of Shem	Gen 11:23
2021		2021+7 3
2023	At age 75 years old, Abram left Haran to the unknown land, then to Egypt, because of the famine.	Gen 12:14 480 x 75 years = 36,000 days
2028		2028+7 4
2030	Melahszte blessed Abram	Gen 14:18-19
2033	Abram married Hagar, the Egyptian.	Gen 16:3
2034	Abram at 86 years old; Hagar bore him a son.	Gen 16:15
2035		2035+7 5
2042		2042+7 6
2047	Yahweh visited Abram; Abram and Sarai names were changed; Abraham, Ishmael and household circumcised; Sodom and Gomorrah burned with the people, except Lot and his two daughters.	Gen 17:5 Gen 17:15 Gen 19: 37 – 38
2048	Isaac born unto Abraham; Abraham circumcised Isaac; Moab and Ammon born unto Lot by his two daughters.	Gen 21:5 Gen 21:4
2049	Sabbatical year	7
2050	Jubilee year	41
2056	Sons and daughters born unto Shem; He died at age 500 years old.	Gen 11:11
2057		2057+7 1
2061	Arphaxas has sons and daughters; He died at age 403 years old.	Gen 11
2064		2064+7 2

2067	Nahor had sons and daughters; Nahor died at age 119 years old.	Gen 11:23	
2071		2071+7	3
2078	Haran died	2078+7	4
2083	Terah died at age 205 years	Gen 11:30	
2085	Sarah died at age 127 years	Gen 23:1-2 2085+7	5
2092		2092+7	6
2099	Sabbatical year		7
2100	Jubilee year		42
2107		2107+7	1
2114		2114+7	2
2108	Birth of Jacob and Esau; Isaac was 60 years old.	Gen 25:26	
2123	Abraham died at age 175 years; and was buried in the cave of Machpelah.	Gen 25:9 480 x 175 years = 84,000 days	3
2126	Sons and daughters born unto Selah; He died at age 408 years.	Gen 11:15	
2128		2128+7	4
2135		2135+7	5
2142		2142+7	6
2147	Esau married Mahalath, daughter of Ishmael.	Gen 28:9	
2149	Sabbatical year		7
2150	Jubilee year		43
2157		2157+7	1
2157	Esau married Adah, a Hittite and Aholibamah the Hivite, both Canaanites; Korah, Gatam, Amalek were the sons of Adah.	Gen 36:2 Gen 36:12	
2164		2164+7	2
2171	Ishmael died at age 137 years. Lunar years, 182.666666		3
2178		2178+7	4
2185		2185+7	5
2192		2192+7	6
2199	Sabbatical year		7
2199	Birth of Joseph		
2200	Jubilee year		44
2207		2207+7	1

2214		2214+7	2
2216	Joseph taken to Egypt at age 17 years old.		
2221		2221+7	3
2228	Death of Isaac	2228+7	4
2229	Joseph appointed governor in Egypt.		
2235		2235+7	5
2237	Famine in Egypt and elsewhere.		
2238	Jacob and generation in Egypt.		
2238	At age 130 years old; Jacob blessed Pharaoh		
2239	Famine ended		
2242		2242+7	6
2249	Sabbatical year		7
2250	Jubilee year		45
2255	Death of Jacob	Embalmed	
2257		2257+7	1
2264		2264+7	2
2271		2271+7	3
2278		2278+7	4
2285		2285+7	5
2292		2292+7	6
2299	Sabbatical year		7
2300	Jubilee year		46
2307		2307+7	1
2308	Joseph died at age 110 years; Embalmed; Joseph saw Ephraim's children of the third generation.	Gen 50:23	
2314		2314+7	2
2321		2321+7	3
2328		2328+7	4
2335		2335+7	5
2342		2342+7	6
2349	Sabbatical year		7
2350	Jubilee year		47
2357		2357+7	1
2364		2364+7	2
2371		2371+7	3
2378		2378+7	4
2385		2385+7	5

2390	The Hyksos was driven out of Egypt; their main weapon was their chariots.		
2392		2392+7	6
2399	Sabbatical year		7
2400	Jubilee year		48
2407		2407+7	1
2414		2414+7	2
2421		2421+7	3
2428		2428+7	4
2435		2435+7	5
2442		2442+7	6
2449	Sabbatical year		7
2450	Jubilee year		49
2452	Queen Hatshepsut was the first woman to be crowned a Pharaoh.		
2457		2457+7	1
2464		2464+7	2
2471		2471+7	3
2478		2478+7	4
2485		2485+7	5
2492		2492+7	6
2499	Sabbatical year		7
2500	Jubilee year		50
2507		2507+7	1
2514		2514+7	2
2521		2521+7	3
2523	Birth of Meriam, sister of Moses and Aaron.	Exodus 2:1 House of Levi; parents: Amran and Jacobi.	
2528		2528+7	4
2532	Birth of Aaron		
2535	Birth of Moses		
2535		2535+7	5
2542		2542+7	6
2549	Sabbatical year		7
2550	Jubilee year		51

2557		2557+7	1
2564		2564+7	2
2571		2571+7	3
2578		2578+7	4
2585		2585+7	5
2592		2592+7	6
2599	Sabbatical year		7
2600	Jubilee year		52
2607		2607+7	1
2614		1614+7	2
2615	Moses/Aaron bid release for the children of Israel in Egypt; first encounter with the pharaoh.	Exodus 5: 1-5	
2616	Amelek fought with Israel in Rephidim;	Exodus 17:8 Numbers 12:1	
2616	Miriam and Aaron spoke against Moses, because he had married an Ethiopian woman; Where is the black Ethiopian Levi family of Moses located today?	Numbers 12:1	
2616	Exodus		
2616	Manna fell for the first time in the second month on the 15th day; Moses lay-away Manna in a secured container, for the future generations. Thunders and lightening on the third day of the third month. Moses brought forward the people to the mountain's base to meet with God; trumpets blasted and He descended with fire on top of Mt. Sinai; Miriam died at age 93 years old; On the twenty third day of the third month; Moses ascended Mt. Sinai for a period of 40 days and 40 nights; Aaron requested golden earrings from the people to make the calf.	Ex 16:32-33 Deut 9:9-11 Ex 31:18 Exodus 32:2-5	
2617	The ark of the covenant was placed in the tabernacle on the first day of the first month, in the second year. On the first day of the second month, in the second year of the exodus, God commanded Moses to take a census of all families of Israel; The count: 603, 550; On the twentieth day of the second month the clouds lifted off the tabernacle of God, and the people journeyed on.	Ex 40:1-2 Num 1:4-6	
2621		2621+7	3

2628		2628+7	4
2635		2635+7	5
2642		2642+7	6
2649	Sabbatical year		7
2650	Jubilee year		53
2655	Death of Aaron at age 123 years in the 40th year, 5th month; Moses last speech to the children of Israel, was on the first day of the 11th month, in the 40th year; Death of Moses at age 120 years, in the land of Moab; Moses was 80 years old at the beginning of the Exodus; Joshua took over leadership role from Moses with the blessing of Yahweh in the year 2655.	Num 33:38 Deut 1:3	
2656	On the 10th day of the first month, Joshua encamped at Gilgal close to Jericho; The younger generations out of the wilderness, were circumcised before participating in the Passover; the children of Israel kept the Passover on the 14th day of the first month at Gilgal. Corn and unleavened bread were eaten; Manna ceased to rain down; Othniel appointed judge for 40 years; An angel of God stood over Joshua with a drawn sword. The Moon and Sun stood still in the battle of Amorites.	Josh 4:19 Josh 5:2-8 Josh 5:10-11 Josh 6:20-21 Josh 5:13, 15 Josh 10:12-13	
2657		2657+7	1
2659	Caleb, at age 85 years old, was given Hebron as an inheritance by Joshua.	Josh 14:13	
2663	Joshua had the children of Israel made a covenant; statutes and ordinances in the land of Shechem, demanding to put away strange Gods and accept Yahweh as the only savior; Joshua died at the age of 110 years old; he was buried in mount Ephraim in the north side of the hill of Gash.	Josh 24:29-30 Judges 2:9	
2664		2664+7	2
2671		2671+7	3
2678		2678+7	4
2685		2685+7	5
2692		2692+7	6
2696	Othniel reign ended		
2696	Judge Ahud reign began; He reigned 80 years.		

2699	Sabbatical year		7
2700	Jubilee year		54
2707		2707+7	1
2714		2714+7	2
2721		2721+7	3
2728		2728+7	4
2735		2735+7	5
2742		2742+7	6
2749	Sabbatical year		7
2750	Jubilee year		55
2757		2757+7	1
2764		2764+7	2
2771		2771+7	3
2776	Ahud reign ended. Deborah reign started. She reigned for 40 years.		
2778		2778+7	4
2785		2785+7	5
2792		2792+7	6
2799	Sabbatical year		7
2800	Jubilee year		56
2807		2807+7	1
2814		2814+7	2
2816	End of Deborah reign. Gideon reign began. He reigned for 40 years.		
2821		2821+7	3
2828		2828+7	4
2835		2835+7	5
2842		2842+7	6
2849	Sabbatical year		7
2850	Jubilee year		57
2856	End of Gideon reign; Abimelech reigned for 3 years.		
2857		2857+7	1
2859	End of Abimelech reign; Tola reigned for 22 years.		
2864		2864+7	2
2871		2871+7	3
2878		2878+7	4
2881	End of Tola reign; Jair reigned for 23 years.		

2885		2885+7	5
2892		2892+7	6
2899	Sabbatical year		7
2900	Jubilee year		58
2904	End of Jair reign; Jephthah reign for 6 years		
2907		2907+7	1
2910	End of Jephthan reign; Ibsan reigned for 10 years		
2914		1914+7	2
2920	End of Ibsan reign; Abdon reigned for 8 years.		
2921		2921+7	3
2928		2928+7	4
2928	End of Abdon reign; Samson reigned for 20 years.		
2935		2935+7	5
2942		2942+7	6
2948	End of Samson reign; Eli reigned for 40 years.		
2949	Sabbatical year		7
2950	Jubilee year		59
2957		2957+7	1
2964		2964+7	2
2968	End of Eli reign		
2971		2971+7	3
2978		2978+7	4
2985		2985+7	5
2992		2992+7	6
2999	Sabbatical year		7
3000	Jubilee year		60
3000	Judge Samuel led the children of Israel within the laws of Moses. He re-adapted the ordinances statues and covenants.		
3007		3007+7	1
3012	End of judge Samuel reign of 12 years. Saul was appointed the first King of Israel. He reigned for 40 years.	Act 13:21 1 Sam 8:5 1 Sam 10:20 – 24	
3014		3014+7	2
3021		3021+7	3
3028		3028+7	4
3035		3035+7	5

3042		3042+7	6
3049	Sabbatical year		7
3050	Jubilee year		61
3052	End of Saul's reign; King David reigned for 40 years.		
3057		3057+7	1
3064		3064+7	2
3071		3071+7	3
3078		3078+8	4
3085		3085+7	5
3092		3092+7	6
3092	Solomon became King at the age of 12 years old.		
3096	Solomon started the temple construction at the age of 16 years old. God appeared to him in a vision.		
3099	Sabbatical year		7
3100	Jubilee year		62
3107		3107+7	1
3112	Temple completed	2Chr 8:1	
3114		3114+7	2
3121		3121+7	3
3128		3128+7	4
3132	End of Solomon reign; he died at age 52 years old; Reheboam reigned for 17 years.	68 years, 3 months; on the Gregorian calendar. 2Chr 10:1	
3135		3135+7	5
3142		3142+7	6
3149	Sabbatical year		7
3149	Abijah reigned for 3 years.		
3150	Jubilee year		63
3152	End of Abijah reign; Asa reigned for 41 years.	1Kings 15:24	
3157		3157+7	1
3164		3164+7	2
3171		3171+7	3
3178		3178+7	4
3185		3185+7	5
3192		3192+7	6

3193	End of Asa reign; Jehoshaphat reigned for 25 years.		
3199	Sabbatical year		7
3200	Jubilee year		64
3207		3207+7	1
3214		3214+7	2
3221		3221+7	3
3228	End of Jehoshaphat reign; Jehoram reigned for 8 years.	3228+7	4
3235		3235+7	5
3236	End of Jehoram reign; Amaziah reigned one year.		
3237	End of Amaziah reign; Athaliah reigned for 6 years. She murdered all the seed of the royal house of Judah, except Joash son of Amaziah who was secretly hidden.		
3242		3242+7	6
3243	End of Athaliah reign.		
3244	Joash reigned for 40 years; he was 7 years old when his reign began.		
3249	Sabbatical year		7
3250	Jubilee year		65
3257		3257+7	1
3264		3264+7	2
3271		3271+7	3
3278		3278+7	4
3284	End of Joash reign; Amaziah reigned for 29 years.		
3285		3285+7	5
3285	Amos predicted the downfall of the Northern kingdom. Amos was the prophet during Amaziah's reign.		
3290	Constantinople was discovered as a Greek colony named Byzantine.		
3292		3292+7	6
3299	Sabbatical year		7
3300	Jubilee year		66
3307		3307+7	1
	Amaziah reconquest of Edom before his reign ended.		
3313	End of Amaziah reign; Uzziah was alive in the day of a terrible earthquake; Azariah (Uzziah) reigned for 52 years.		
3314		3314+7	2

3321		3321+7	3
3328		3328+7	4
3335		3335+7	5
3342		3342+7	6
3349	End of the ancient world.		
3349	Sabbatical year		7
3350	Jubilee year		67
3357		3357+7	1
3364		3364+7	2
3365	End of Azariah reign; Jotham reigned for 16 years.		
3371		3371+7	3
3378		3378+7	4
3381	End of Jotham reign; Ahaz reigned for 16 years. He committed atrocities in the temple of Yahweh; he cut up the temple vessels, and he constructed altars unto foreign Gods.		
3385	Isaiah prophesied that Babylon would fall.	3385+7	5
3392		3392+7	6
3397	End of Ahaz reign; Hezekiah reigned for 29 years.		
3399	Sabbatical year		7
3400	Jubilee year		68
3405	Death of Buddha		
3407		3407+7	1
3414		3414+7	2
3421		3421+7	3
3426	End of Hezekiah reign. In the 4th year of his reign, 340 Assyrians attacked the Northern kingdom and dispersed the people. In his first month of reign, he repaired the temple of God; he restored the vessels; he kept Passover in the second month on the 14th day, because the people were not prepared; Isaiah was present; Hezekiah's life was extended 15 years by God.		
3426	Manasseh reigned for 55 years		
3428		3428+7	4
3435		3435+7	5
3442		3442+7	6
3449	Sabbatical year		7

3450	Buddhism emerged in the 6ᵗʰ century BC, in India.		
3450	Jubilee year		69
3456	King Darius, the Persian, crushed the revolt of the Ionian Greek cities.		
3457		3457+7	1
3464		3464+7	2
3471	The Greek's final threat from the Persians. An all out victory.		3
3478		3478+7	4
3481	End of Manasseh reign. During his reign as king, he arranged Isaiah's death; Amon reigned for 2 years.		
3483	End of Amon's reign; Josiah became king. He reigned 31 years.		
3485		3485+7	5
3492		3492+7	6
3496	Daniel prophesied about Yehshua.		
3499	Sabbatical year		7
3500	Jubilee year		70
3507		3507+7	1
3514		3614+7	2
3514	End of Josiah reign. He was wounded in battle by an Arrow and later died. Josiah kept a record Passover, as per Mosaic laws. He followed the Prophetic calendar with the appropriate dates of appointment; he purged the temple. Ezekiel foretold the return of the children of Israel from Babylon. Jehoahaz reigned for three months. The king of Egypt Necho the Pharaoh made Eliakim king, and changed his name to Jehoiakim; He reigned 11 years.		
3521		3521+7	3
3521	Jeremiah and Baruch gave Jehoiakim a scroll to read. He shredded it and threw it in the fire place.	Jeremiah 36: 22-23 The king sat in the winter house in the 9ᵗʰ month.	
3525	Nebuchadnezzar killed Jehoiakim and he made Jehoichin king, but later replaced him with Zedekia (Mattaniah). He reigned 11 years. He was the uncle of Jehoichin.	Jer 52:1 2Chr 36:16	

3528		3528+7	4
3533	Zedekia broke his pledge with Nebuchadnezzar.		
3535		3535+7	5
3536	End of Zedekia's reign. His two sons killed. Zedekiah blinded and taken to Babylon. Jehoichin King of Judah was taken captive, along with his family, Ezekiel; Asalad. Judean leaders and craftsmen; the princes of Judah in Riblah were killed.		
3542		3542+7	6
3542	Athens defeated by the Spartans.		
3544	In the 5th month, on the 7th day, in the 19th year of the reign of King Nebuchadnezzar of Babylon his captain of his guard Nebuzaradan burned the temple of God, and the houses in Jerusalem. The residents were exiled to Babylon; but some remnants remained to work the field. The treasures were taken from the temple to Babylon.	Ref: Jeremiah 52:12 States that in the 5th month, on the 10th day Nebuzaradan burned the temple in the 19th year. (Conflicting dates) 2 Kings 25:8-12	
3549	Sabbatical year		7
3550	Jubilee year		71
3557	Governor Gedaliah was murdered.	3557+7	1
3562	Jehoichin was released from prison by Evil – Merodach son of Nebuchadnezzar who became King; Nebuchadnezzar died.	2 Kings 25:27-30	
3564	Belshazzar succeeded Evil – Merodach. He reigned for seventeen years.	3564+7	2
3571		3571+7	3
3578		3578+7	4
3585		3585+7	5
3592		3592+7	6
3596	Cyrus the great conquered the city of Babylon at the age of 62 years old. He had the people of Israel returned back to their homeland in Jerusalem, after 52 years in exile; 42,360 residents of Judah and Benjamin; 7,337 male and female servants; 245 singers; it took four months to journey from Babylon to Jerusalem.		

3597	Zerubbabel built an altar and made offerings unto God.		
3598	In the second year out of captivity in the second month; Zerubbabel; the priests; the Levites; the remnants and Jeshua commenced working on the second temple. The king of Persia Artaxerxes, Ahasuerus in the beginning of his reign stopped the construction process.	Ezra 3:8	
3599	Sabbatical year		7
3600	Jubilee year		72
3600	Darius made a decree to resume the building of the temple.		
3607		3607+7	1
3610-3614	Egypt was conquered by the Greek.		2
3617	Shimon Hatzadik met Alexander the great; Alexander the great conquered Palestine.		
3618	Greek king Alexander the great came under the rule of Ptolomies.		
3620	Second temple completed.		
3621		3621+7	3
3624	A squadron of 200 elephants fought in battle of the Hydaspes by Alexander defeating the Indian King Borus.	The elephants were the war tanks in those days.	
3628	Aristotle was the most influential of the Greek philosophers.		4
3633	Ezra, the priest read the mosaic laws to a congregation of men and women on the first day of the 7th month; all the people made booths, and sat inside; in the 24th of the 7th month the fasted; in exile they adapted the Babylonian's culture and their calendar.		
3635		3635+7	5
3642		3642+7	6
3649	Sabbatical year		7
3650	Jubilee year		73
3657		3657+7	1
3664		3664+7	2
3671		3671+7	3
3678		3678+7	4

3685		3685+7	5
3686	Rome invaded Sicily.	264 BCE	
3692		3692+7	6
3699	Sabbatical year		7
3700	Jubilee year		74
3707		3707+7	1
3714		3714+7	2
3721		3721+7	3
3728		3728+7	4
3729	Chin Chi Wongti unified China and proclaimed himself first emperor of China. He reigned 12 years, and linked miles of old walls as a defense strategy against the Hans, a wall of national identity.		
3735		3735+7	5
3742		3742+7	6
3749	Sabbatical year		7
3750	Jubilee year		75
3750	The ancient Mayans made their appearance in Mexico; they were versed in astronomy.		
3757		3757+7	1
3764		3764+7	2
3771		3771+7	3
3778		3778+7	4
3785		3785+7	5
3786	Syrian ruler Antiochus IV attempted to destroy all scrolls of the laws, found in Palestine.		
3792		3992+7	6
3799	Sabbatical year		7
3800	Jubilee year		76
3807		3807+7	1
3811	The Maccabees rebelled against the Greek and Syrian rule.		
3814		3814+7	2
3821		3821+7	3
3828		3828+7	4
3835		3835+7	5
3842		3842+7	6

3845	Paper was invented in China.		
3849	Sabbatical year		7
3850	Jubilee year		77
3857		3857+7	1
3864		3864+7	2
3871		3871+7	3
3873	The astronomical calendar was written.		
3878		3878+7	4
3879	Rome's dominion over the Mediterranean world; Rome crushed Spartacus revolt.		
3885		3885+7	5
3887	Roman's conquered Palestine, and created the province of Judea.		
3888	The Romans conquered Egypt.		
3892		3892+7	6
3895	The Romans invaded Scotland.		
2899	Sabbatical year		7
3900	Jubilee year		78
3905	Julius Caesar introduced the Julian calendar (45 BC).		
3907		3907+7	1
3908	The battle in ancient Philippe; a civil war between Romans after the assassination of Julius Caesar.		
3913	Cleopatra financed Anthony's military in Persia campaign.		
3914		3914+7	2
3919	The Roman senate stripped Anthony of his political offices and declared war on Cleopatra.		
3921		3921+7	3
3928		3928+7	4
3935		3935+7	5
3942		3942+7	6
3947	The first known translation of the Hebrew scriptures was the Septuagint version in Greek. Greek outside Palestine who spoke Hebrew was used to do translation. .		
3949	Sabbatical year		7

3949	Yehshua was born in a manger of a livestock.	In the beginning of the 188th moon cycle of a 21 years prophetic cycle, on 13th day of the 1st month, was a new moon on Passover.
3950	Jubilee year	79
3955	Palestine came under direct Roman rule.	
3957		3957+7 1
3961	Age 12 years old, Yehshua and parents returned from Egypt; after the death of King Herod.	
3963	Tiberius reigned for 23 years.	
3964		3964+7 2
3971		3971+7 3
3976	Yehshua (Jesus) was baptized by John the Baptist.	Yehshua's ministry started when John was arrested.
3978		3978+7 4
3979	Yehshua predicted the destruction of the second temple; He stopped all animal sacrificed, and money exchange transacted in the temple of God. Yehshua's final Passover, in the evening of the 13th day of the first month. He did not participate in the drinking of wine, because He kept his vow as a Nazarene, so was Samson, the judge, and John the Baptist who did not cut their hairs; Yehshua was crucified after the Passover on the 13th at 9 pm; the 15th day of the first month in the hour of darkness He arose from the dead. Later He was witness by many, and spoke to His disciples during His 40 days travelling around; Act 1:3; 1Cor 15:5-7 A cloud received Him and He went out of sight. Act 1:9-11; Luke 21:27; John 14:3; Mark 16:21-19 Matthew 27:50-53 When Yehshua was crucified, He cried out aloud and gave up the ghost. The earth quaked; the grave opened up and many bodies of the righteous that slept in the graves arose; this took place after His resurrection.	It was a new moon when Yehshua had his last Passover; it was the 10th year of the 189th moon cycle in the year 3979, in the first month; note: Every twenty one years is a prophetic cycle; Yehshua was born and died on the same day of Passover.

3985		3985+7	5
3986	Tiberius reign ended in		
3992		3992+7	6
	St Peters church stands in Rome, where one of Yehshua's twelve disciples was martyred and crucified upside-down. His tomb is places under the altar of the church.		
3999	Sabbatical year		
4000	Jubilee year		80
4003	Nero, at age 16 years old succeeded his stepfather Claudius for the Imperial throne.		
4004	Anarchy and revolt throughout Judea.		
4007			
4009 - 4011	Second temple destruction. Abomination and desolation committed in the temple. Daily sacrifices were taken away. It was 1290 days, or 2 years, 6 months and 8 days.	Daniel 12:11	
4049	Sabbatical year		
4050	Jubilee year		81
4099	Sabbatical year		
4100	Jubilee year		82
4149	Sabbatical year		
4150	Jubilee year		83
4199	Sabbatical year		
4200	Jubilee year		84
4249	Sabbatical year		
4250	Jubilee year		85
4299	Sabbatical year		
4300	Jubilee year		86
4349	Sabbatical year		
4350	Jubilee year		87
4399	Sabbatical year		
4400	Jubilee year		88
4449	Sabbatical year		
4450	Jubilee year		89
4499	Sabbatical year		
4500			90
4549	Sabbatical year		

4550	Jubilee year	91
4599	Sabbatical year	
4600	Jubilee year	92
4649	Sabbatical year	
4650	Jubilee year	93
4699	Sabbatical year	
4700	Jubilee year	94
4749	Sabbatical year	
4750	Jubilee year	95
4799	Sabbatical year	
4800	Jubilee year	96
4849	Sabbatical year	
4850	Jubilee year	97
4899	Sabbatical year	
4900	Jubilee year	98
4949	Sabbatical year	
4950	Jubilee year	99
4999	Sabbatical year	
5000	Jubilee year	100
5049	Sabbatical year	
5050	Jubilee year	101
5099	Sabbatical year	
5100	Jubilee year	102
5149	Sabbatical year	
5150	Jubilee year	103
5199	Sabbatical year	
5200	Jubilee year	104
5249	Sabbatical year	
5250	Jubilee year	105
5299	Sabbatical year	
5300	Jubilee year	106
5349	Sabbatical year	
5350	Jubilee year	107
5399	Sabbatical year	
5400	Jubilee year	108
5449	Sabbatical year	
5450	Jubilee year	109

5499	Sabbatical year	
5500	Jubilee year. I will bring forward a seed out of Jacob and out of Judah an inheritor of my mountains; and mine elect shall inherit it and my servants shall dwell there. Isaiah 65:9	110
5549	Sabbatical year	
5550	Jubilee year	111
5599	Sabbatical year	
5600	Jubilee year	112
5649	Sabbatical year	
5650	Jubilee year	113
5699	Sabbatical year	
5700	Jubilee year	114
5749	Sabbatical year	
5750	Jubilee year	115
5799	Second industrial revolution.	
5799	Sabbatical year	
5800	Jubilee year	116
5849	Sabbatical year	
5850	Jubilee year	117
5899	Sabbatical year	
5900	Jubilee year	118
5949	Sabbatical year	
5950	Jubilee year	119
5992	Michael, the fifth angel of God shall receive the key for the bottomless pit Rev 9;1 Satan shall be detained in the bottomless pit for 1000 years. Rev 20:1-3	Revelation 20:4. The chosen ones will reign and live with Yehshua for 1000 years.
5999	Sabbatical year	
6000	Jubilee year	120
6049	Sabbatical year	
6050	Jubilee year	121
6099	Sabbatical year	
6100	Jubilee year	122
6149	Sabbatical year	

6150	Jubilee year		123
6199	Sabbatical year		
6200	Jubilee year		124
6249	Sabbatical year		
6250	Jubilee year		125
6299	Sabbatical year		
6300	Jubilee year		126
6349	Sabbatical year		
6350	Jubilee year		127
6399	Sabbatical year		
6400	Jubilee year		128
6449	Sabbatical year		
6450	Jubilee year		129
6500	Jubilee year		130
6549	Sabbatical year		
6550	Jubilee year		131
6599	Sabbatical year		
6600	Jubilee year		132
6649	Sabbatical year		
6650	Jubilee year		133
6699	Sabbatical year		
6700	Jubilee year		134
6749	Sabbatical year		
6750	Jubilee year		135
6799	Sabbatical year		
6800	Jubilee year		136
6849	Sabbatical year		
6850	Jubilee year		137
6899	Sabbatical year		
6900	Jubilee year		138
6949	Sabbatical year		
6950	Jubilee year		139
6957			
6964			
6971			
6978			
6985		6985+7	5

6986	God shall give two witness authority and special powers; they will walk Jerusalem for two and a half years (Prophetic calendar of 480 days) then will be killed in Jerusalem and taken up. The entire world shall witness through main stream media.	Revelation 11:2-12 Haggai 1:1; 2:2, 23 The two mysterious witnesses are: Joshua, the son of Josedech, the High Priest, and Zerubbabel, the son of Shealtiel, the governor of Judah. Zerubbabel shall bring the new temple Headstone from heaven to earth. Zech 4:7, 9:12,14; Revelation 11:11-13
6990	More tribulations: Locust plague; fire; sulfur; pestilence; hailstorm; heavy rainfall.	Ezekiel 38: 22 Ezekiel 39:6 Zechariah 6:1-6
6992		6992+7 6
6992 Ezekiel 38:15, 16, 18 Zechariah 5:2-4	Satan is released for a short time. He will nominate a president *Shem# 666 and Serug a sixth generation after Shem.* They shall muster nations, (*MAGOG*) to do battle with Jerusalem with the chosen ones. *MAGOG* attacks from the north in full strength; A technological warfare against the angels of God with thousands of chariots (UFO). Psalm 68:17, 33. A counter attack: fire sent down from heaven and destroyed *GOG* and *MAGOG*, that came against Jerusalem; also, previous nations that attacked Jerusalem, and plundered her shall be eradicated; measure for measure. Tribulation intensifies for seven years.	*All electronic systems on planet and in space shall be disabled by God's angels in their chariots, prior the final battle between good and evil.* Zechariah 12:9

6993 Isaiah 34:4-9 Isaiah 42:9 Last 21 year Prophetic cycle of 333 (333x21=6993) New creation in 7000 of 21 year Prophetic cycle = 7000÷21=333.33333 Ezekiel 39:11-12 Amos 9:2-4,9 Psalms 139:8 Jeremiah 23:24 Isaiah 53:5-12 Zechariah 14:4 Ezekiel 38:20 Isaiah 49:11 7 Mountains of Jerusalem Revelation 17:9	The moon shall be for the new moon celebration, and it shall not shine. The other hosts of heaven, the stars and sun shall fall; After the war of *GOG* and *MAGOG* their shall be a clean-up operation of the dead bodies for seven years; a seven-month operation to bury the dead in the valley of *Hamon-GOG*; burial officials shall place markers at every skeleton found in the field. All the enemies of Jerusalem flesh shall dissolve, while they are standing; death shall come to all of Esau's generation. Heaven shall dissolve with all its hosts. Yehshua shall appear in the clouds; Leviathan shall be destroyed; Terrible earthquakes on the entire earth; Mount of olives shall be split; large valley formed. Yehshua shall stand on Mount of olives. Tribulation shall not cease until Yehshua reigns; hills and mountains. Tall buildings and all tall trees shall be brought down. Jerusalem will be on a mountain of a flat plain between six other mountains. The tribes shall be surrounding the temple of God.	Isaiah 45:8 Zech 14:12 Is 27:1 Is. 47:13-14 Isaiah 42:15 Isaiah 54:10 Revelation 13:1 Mount Olives
6994 New Heavens and a new earth.	New heavens and a new earth are created. The dead in righteousness will come to life. Yehshua's first gathering of the exiled tribes. The righteous shall be cleansed by refining through fire. Earth shall be destroyed by fire, in order to purge the land from various wastes and pollution.	Earth Destroyed Isaiah 64:2 Ezekiel 37:25-27 Zechariah 13:8-9 Amos 9:9-14 Revelation 18:9-24
6995	In- gathering of the remaining tribes from Assyria and Egypt.	Ezekiel 37:12 Hosea 13:14 Ezekiel 47:22 Amos 9:15
6996	Yehshua shall sit on the throne and judge. The devil and his angels shall be thrown in the lake of fire.	Isaiah 16:5 Ezekiel 37:22-24
6997 The book of all souls	The book shall be opened; Judgment on the living and the unrighteous shall arise from the graves. The unrighteous shall be thrown in the lake of fire.	

6998	Division of land among the tribes (14)	Ezekiel 36:8, 10, 17, 24, 28, 34. Daniel prophesied in chapter 12:12. "Blessed is he that waits and reaches 1335 days". It is exactly two years from 6998.
6999	Sabbatical year	7
6999	A new Jerusalem established. Handing over of refined ones. Also a new kingdom of priests to Yahweh.	Isaiah 65:17
7000	A pure language shall be spoken.	140th Jubilee year.
7000	Jubilee and joyful year; grand celebration with Yahweh and chosen ones.	Daniel 12:12. Prophesy came to pass pertaining to 1335 days.
7000	A new song of Jerusalem will be sung; The song is already known by his chosen ones. No unclean or uncircumcised shall enter Jerusalem. All nations shall come up to Jerusalem to celebrate Sabbaths and new moon; No more warfare.	Isaiah 65:14 Isaiah 56:7 Isaiah 66:23

Isaiah 66:23. Nations shall worship the lord from new moon to new moon, and from Sabbath to Sabbath. Zechariah 14:16.

CHAPTER SEVEN

BIBLICAL HISTORY:

The reintroduced Prophetic calendar shall reveal to the world, that the people are younger in age than previously recorded on their birth certificates and other documents. In fact, everything that was given an age in the universe by educators and other professionals are lesser in age than previously thought. The re- adaptation of the Prophetic calendar would call for modification of books, banking laws, international affairs or countless items, and in many other areas to be in synchronization with the other worlds; and with the nations on planet earth. The calendar is not required to be incorporated with a lunar calendar, Solar or Astronomical calendars. There is no need to add days or a month in keeping synchronized with the two seasons, or to maintain consistency with programs, and celebrations of religious observances or other worldwide events. If adapted, the programs of changing or by adding to prevent seasonal drift (global warming) can be resolved, and it will be very simple to compute on a year to year basis - only the numbers in a new year will be required changing. God spoke to Moses and Aaron in the land of Africa, and he told them that the beginning of months of the year is the month of Abib, the first month. *Exodus 13:4; 23:15; 34:18; Deuteronomy 16:1.*

The Gregorian calendar is the calendar utilized by many countries today, and it is not synchronizing with the cycle of nature or the other worlds.

MONTHS OF FORTY DAYS:

In using the biblical text in defining the Prophetic calendar of the Lord, here is some scripture verses reinforcing, that the true calendar carries forty days in each of its twelve months:

- **Numbers 11:19-20 (A clue)**
  You shall not eat:
  One day or two days
  Five days or ten days
  Twenty days or even a month (40 days)
- **Numbers 14:34.** After the numbers of days in which you searched the land; even forty days and each day for a year….'**Exodus 24:18; 34:28; Deuteronomy 9:9; 10:10 Moses was in the mountain for forty days and forty nights.**
- The rain was upon the earth for forty days and forty nights. **Genesis 7:12**
- The Egyptians embalmed the body of Jacob in Egypt and it took forty days in the process. Joseph gave that command to the Egyptian physicians. Also Joseph body was embalmed when he died in Egypt. **Genesis 50:3, 26.**
- Ezekiel lay on his right side to bear the sins of the house of Judah for a period of forty days. **Ezekiel 4:6.**
- **Numbers 14:34.** Because of the false rumors of the ten spies that were sent out on a forty days mission to spy out the land; the children of Israel was punished for forty years (A day per year).
- **Leviticus 12:2-4; Luke 2:22.** Women who bear a male child are unclean for seven days. Their purification is thirty three days; 33 days + 7 days = 40 days or one month.
- It takes a period of forty days for a fetus to form in a woman, and 7x40 days for a new born (280 days) or seven months.
- **Jonah 3:4** Jonah said, "Yet forty days and Nineveh shall be overthrown."
- **1Kings 19:8** Elijah journeyed forty days and forty nights unto Mount Horeb, the mountain of Yahweh.
- **Acts 1:3** After Yehshua (Jesus) had risen from the dead; he was witnessed by many and continued to perform miracles for forty days in the land.
- **Deuteronomy 1:3** In the fortieth year of the eleventh month, on the first day of the month (A day per year), Moses spoke to the children of Israel according to the commandment that God had given to him.
- **Genesis 8:5; 8:7.** In the tenth month on the first day of the month -on top of the mountains were visible. At the end of forty days (one month) Noah sent out a raven.
- **Matthew 4:2** Yehshua had fasted forty days and forty nights in the wilderness.

THE FLOOD YEARS – 1656-1657 (GENESIS 7:6, 11; 8:14):

- Flood started: 17th day of the second month in the year 1656 (Noah's age 600 years old), Genesis 7:6, 11.
- Rain and waters from beneath: 40 days and 40 nights Genesis 7:11, 12.
- Water prevailed on Earth: 150 days Genesis 7:24.
- Water abated from the Earth: 150 days Genesis 8:3
- Ark rested on Mount Ararat on the 17th day of the seventh month in the year 600 Genesis 8:14.
- Mountain top visible on the 1st day of the tenth month, Genesis 8:5.
- Window of Ark opened at the end of 40 days Genesis 8:6.
- Noah released dove on the 7th days, Genesis 8:8, 9.
- Second dove sent out, 7th days Genesis 8:10, 11.
- Third dove released days7, Genesis8:12.
- Water dried up from the Earth the 1st day of the first month. Genesis 8:13.
- Earth totally dried up on the 27th day of the second month. Genesis 8:14.
- Total amount of days on dry land were 40 days + 27 days = 67 days.
- Time the Ark was on water and land from the year 1656-1657 is 423 days on water and 67 days on land equaling 490 days. From the 17th of the second month, year 1656 to the 27th of second month, year 1657= 490 days.

* Note that Noah's birthday fell on the 17th day of the second Month, and the family touched down on solid ground on the 27th day of the second Month. It appears that Noah celebrated his birthday aboard the Ark.

COUNTDOWN FROM THE TRIBULATION PERIOD

Tribulations of the various kinds will intensify as the souls from heaven will be sent to earth. A quickening increase of the world population will be taking place in the end times, while some souls will be subsequently evoked from their graves. Mankind shall endure hardship for a short period of time in the latter days. Difficult time periods will last for two hundred days (five months) on the Prophetic calendar. All saints and the chosen ones lying in their graves will be the first to be resurrected from the dead, and to dwell with the Messiah for one thousand years. The rest of the dead will be resurrected after the release of Satan.

In essence, the tribulation period started when the second temple was destroyed by the Romans in the year 4009. Beginning in the year 4475 to date, many deaths are caused by severe natural disasters, accidents, illnesses and wars. From the time of destruction of the first temple in the year 3544 by the Babylonian king Nebuchadnezzar, to the second temple destruction in 4009, there were no natural disaster reports available or recorded for the period of four hundred and sixty five years (465 years). Tribulation works in conjunction with the sun and moon. No scientific venture or experiments can prevent nature from taking its course. No one can accurately give an exact date and time of a disasters as earthquakes, tidal waves or other tribulations in the future, unless it is prophetic.

The Jubilee Calendar is working in conjunction with the Prophetic calendar, which is extremely accurate in pinpointing dates of disasters - either scientifically or prophetical. When a word comes from above, God will give a date on the calendar in use today. In most events He will give dreams to those that have the spirit of God in them, but not to those that have the spirit of Lucifer.

There are some notable disasters as storms, earthquakes, tidal waves or hurricanes. Those disasters will come within a twenty one year cycle or within 10,080 days or if directed by God. A natural disaster may either cause fatalities, calamities, homelessness, blackouts, and will cause many to suffer, or may be lacking in the basic essentials as water and food, in order to sustain life. The many organizations around the world as the churches, temples, synagogues, Salvation Army and other non- profit organizations, should be permanently committing their members by getting them totally involved in rendering some assistance to those desperately in need. Most victims may be affiliated with one of the organizations, and most of them are contributing ten percent gross of their earnings, on a monthly basis to those organizations. The victims will be looking forward to some assistance, support, and some love from their particular organization.

THE MODERN WORLD:

Mankind has invented new sophisticated technological equipment's to restructure, manipulate and alter nature to modify the weather, in order to artificially produce rainfall out of its season. It was published that China had modified its weather, ensuring that there was no hindrance to the Olympics that was hosted by their nation.

- Leviticus 19:19. God has emphasized that man or mankind should not mix their cattle genders with a diverse kind. Today this commandment has been willfully violated. They have gone to the extent of cross pollinating plants and seeds of the various kinds.
- There is a growing fear and predictions of natural disasters and terrorists activities worldwide as the end of the age is coming to a close. The scriptures describe those tribulations as birth pains; but each may last for a short time.
- Tribulations shall be accelerating by a three and a half years period from the year

2011, using that year as a marker to the continuing tribulations. Psalm 68; 17, 33; *Deuteronomy 33; 2, states that the Almighty God has in his possession 20,000 chariots of fire (UFO's) in His Kingdom.* They will be used in the last days to do battle on planet earth. Isaiah 66:15; 9:5. The chariots seen by many all over the world are known as UFO.

- At the ending of the age, those with the dormant spirit of Lucifer shall then be activated by the Almighty God. The world will be witnessing an increase of evil- doings by those that have the spirit of Lucifer, and family feud will be on the rampant. Brother will hate brother. Parents will give up their children to the authorities of the beast, who will be in control of many nations. Within a family of ten people, eight family members can have the spirit of Lucifer.

To reiterate, there is three and a half Prophetic years. The conversions to the Gregorian calendar years are four years and six months, or 1,680 days. These days will be targeted to the year 2012 through 2014, a period of further tribulations with natural disasters and portent things happening in the Cosmos.

FOUR HUNDRED YEARS OF OPPRESSION

God's chosen people were in Egypt under oppression for four hundred years. They were oppressed by a new Pharaoh, who did not know Joseph as the previous Pharaoh did in the days of the famine, in the year 2238 to 2240. This was foretold to Abram by God, before his name was changed to Abraham in the year 2047. God told Abram that his offspring shall be foreigner in a country not their own. They will be slaves, and will be afflicted for four hundred years. **Genesis 15:13; Acts 7:6; Exodus 1:11-14.**

It was a total of four hundred and thirty years from the time Abraham was in Egypt to the time Joseph and the generation of Jacob spent in the land of Egypt, until the exodus in 2616.

TIMELINE FROM NOAH TO JACOB'S GENERATION

Event	From	To	Years
Birth/ Death of Noah	1056	2006	950
Flood Years	1656	1657	1
Birth/ Death of Shem	1556	2056	500
Birth/ Death of Eber	1723	2187	430
Birth/ Death of Peleg in the day of the division of the earth	1757	1967	210
Birth/ Death of Terah	1878	2083	205
Birth/ Death Abraham	1948	2123	175

Event	From	To	Years
Birth/ Death of Sarah	1958	2085	127
Birth/ Death of Ishmael	2035	2172	137
Birth/ Death of Isaac	2048	2228	180
Birth/ Death of Jacob	2108	2255	147
Birth/ Death of Joseph	2198	2308	110
Joseph, sold to Egypt	2215		17
Joseph, appointed governor	2228		30
Jacob, and Generation moved to Egypt; at the age of 130 years old, he blessed Pharaoh	2238	2255	17
Famine lasted	2238	2240	2
Moses/ Aaron/ Bid, for the release of the children of Israel from the pharaoh	2615	2616	1
Birth/ Death of Meriam	2523	2616	93
Birth/ Death of Aaron	2532	2655	123
Birth/ Death of Moses	2535	2655	120
Saul, as first king of Israel (Acts 13:21)	3012	3052	40
Birth of king David, on to his reign.	3022	3052	30
Reign of king David	3052	3092	40
Birth of king Solomon, on to his reign	3080	3092	12
Reign of king Solomon	3092	3132	40
From the Exodus to first temple construction	2616	3096	480
Second temple construction	3598	3620	22
Birth of Yehshua (Jesus)/ Death	3949	3979	30
Prediction made by Yehshua pertaining to destruction of the second temple.	3979	4009	30
First temple destruction, to second temple destruction	3544	4009	465
First temple construction, to its final destruction	3096	3544	448
From Abraham time spent in Egypt; Joseph's time, and Jacob's generation in Egypt. Romans 6:14.	30	400	430

Those who are led in the spirit of Yehshua are not under the old law. Galatians 3:10 said that the followers under the works of the law are under the curse. Curse is everyone that discontinues not in all things that was written in the old covenant. A typical example is in Roman 2:25 which states that circumcision is truly authentic, if only you maintain the entire old law. But if you are a breaker of that law your circumcision is made uncircumcision.1 Corinthians 7:19.

BIBLICAL PROPHETIC CALENDAR

FIRST MONTH

Day 1	Day 2	Day 3	Day4	Day5	Day 6	Day 7
Gen1:3 Light created **Gen 8:13** In the year 601 flood waters dried up. **Ex 40:2** Tabernacle erected **Ezek 45:18 -20** Sanctuary cleansed **Num 20:1** Meriam died　1	**Ezek 43:22** Sanctuary altar cleansed **Gen 1:6** Division of waters　2	Dry land appeared; seas separated from the land. Herbs; grass; fruit trees **Gen 1:11**　3	New moon; stars; sun, created **Gen 1.14-1**　4	Birds; fishes; leviathan; beasts, cattle; creeping things created **Gen 1:20**　5	Adam was created and a woman was created **Gen 1:26-28**　6	God ended his work and rested. Blessed and sanctified it. **Gen 2:2**　7
2 Chr 29:17 Sanctification of temple　8	**Gen 20:11** Moses hit the rock twice and water came out. (2616)　9	Passover Lamb kept **Ezek 43:4** God's esteem in the temple, by east gate　10	**Josh 5:23** Circumcision of new generation (2656)　11	12	*Yehshua and Disciples at Passover. Matthew 26:20 MARK 15:3 Yehshua died at 9::00 pm　13	**Gen 15:13** 400 years oppression; **Ex 12:41** 430 years;　14
Unleavened bread Yehshua, arose from the dead, on the third day. *Easter Sunday.　15	**Josh 5:10** Manna stopped (2656)　16	UB　17	UB　18	UB　19	UB *Good Friday　20	UB　21
No work **Lev 23:10 -12** Priest waves sheaf after Sabbath offerings　22	23	**Gen 2:5,6** Herbs were already in the ground, but no rain for it to sprout; rain came.　24	25	26	27	28
Gen 2:11 The Pison Gihon, Hiddekel Euphrates, was created in the summer.　29	30	31	32	33	34	35
36	37	38	Blow trumpet. Beginning of new months **Num 10:10**　39	40		

SECOND MONTH

Day	Day	Day	Day	Day	Day	Day
Num 1:1 Census taken in 2nd year (2617) 1	**1 Kings 6:1** 480 years out of Egypt temple of Solomon **2 chr 3:2** (3096) 2	3	4	5	6	7
8	9	10	11	12	13	**Num 9:11** Second Passover. 14
Ex 16:13 Quails **Ex 16:1** Arrived in the wilderness (2616) 15	**Ex 16:14** Manna fell (2616) 16	**Gen 7:11** Flood year 600; 40 days and 40 nights (1656) 17	18	19	**Ex 16:32,33** Manna in container for safe keeping for future generations **Num 10:11; 9:17** Clouds lifted off tabernacle 20	21
22	**Num 11:1** Fire consumed the complainers in the wilderness (2616) 23	Day of Pentecost 24	25	26	**Gen 8:14** Earth fully dried after flood waters (1657) 27	28
29	30	31	32	33	34	35
36	37	38	39	40		

THIRD MONTH

Day	Day	Day	Day	Day	Day	Day
1	2	3	4	5	6	7
8	9	**Baruch 1:8,9** Vessels taken from temple. By Nebuchadnezzar (3544) 10	11	12	13	14
Moses went up unto mount Sinai to receive instructions **Ex 19:1** (2616) 15	Moses ascended mountain. Clouds covered mountain for six days **Ex 24:16** Preparations to receive commandments **Ex 19: 10-12** 16	40 days 40 nights flood waters increased (1656) 17	**Ex 19:16** Thunders and lightening at mount Sinai. Earthquake followed (2616) 18	19	20	21
22	**Ex 24:18** On the 7th day Yahweh called Moses on mount Sinai for 40 days and 40 nights 23	24	25	26	27	28
29	30	31	32	33	34	35
36	37	38	39	40		

Fourth Month

Day	Day	Day	Day	Day	Day	Day
				Ezk 1:4 - 28 UFO (3541) 5		
1	2	3	4		6	7
8	9	10	11	12	13	14
15	16	17	18	19	20	21
Deut 9:14 Yahweh told Moses that he shall erase the people, and make him an Almightier nation than they. 22	**Ex 32:19** Moses spent 40 days and 40 nights: Moses descended; golden calf; broken tablets. 23	On this day Moses prayed to win atonement for the people who sinned with the golden calf **Ex 32:30** (2616) 24	25	26	27	28
29	30	31	32	33	34	35
36	37	38	39	40		

FIFTH MONTH

Day	Day	Day	Day	Day	Day	Day
Ezra 6:9,10 Ezra taught laws, statutes and judgments to the people of Israel. (3633) **Deut 10:6** Aaron died on mount Hor at age 123 years. (2655) 1	2	3	4	5	6	7
8	9	**Jer 52:12,13** Nebuzaradan burned the temple in the 19th year (3644) 10	11	12	13	14
15	16	17	18	19	20	21
22	**Ex 34:4** Moses carved two new tablets and ascended mount Sinai. 13 attributes **Ex 34:5-7** 23	24	25	Jacob and family entered Egypt. (2238) 26	27	28
29	Mourning ends for Aaron 30	31	32	33	34	35
36	37	38	39	40		

SIXTH MONTH

Day	Day	Day	Day	Day	Day	Day
Haggai 1:1-11 Zerubbabel was told to re-build temple 1	2	3	4	**Ezek 8:1** Idolatry in the Temple in the 6th year. (3542) 5	6	7
8	9	10	11	12	13	14
15	16	17	18	19	20	21
Moses descended after 40 days and 40 nights with two tablets; face radiated. 22	23	**Haggai 1:14,15** Joshua and the people worked building the temple, in the second year of king Darius (3600) 24	25	26	27	28
29	30	31	32	33	34	35
36	37	38	39	40		

SEVENTH MONTH

Day	Day	Day	Day	Day	Day	Day
Num 29:1 **Day of Trumpets** **Neh 8:2-14** Nehemiah read laws. 1	**Neh 8:13** Ezra addressed priests and Levites. (3633) 2	3	4	5	**Gen 8:3** Water abated at end of 150 days (1656) 6	7
8	9	**Lev 16:29,30** Day of atonement; fast. **Lev 25:8-9** trumpet in Jubilee year. 10	**Ex 25:8** God promised to live with his people Israel. **1 kings 6:13** 11	12	13	**Lev 23:34** Eve of booths 14
Lev 23:35 Rest day booths 15	Booths 16	**Gen 8:4** Ark rested on mount Ararat. Booths (1657) 17	Booths 18	Booths 19	Booths 20	21
Lev 23;39 Rest day 22	**1Kings 8:12-14** Solomon dedicated temple **2 Chr 7:10** (3112) 23	**Neh 9:1-3** Confession; Fast day (3633) 24	25	26	27	28
29	30	31	32	33	34	35
36	37	38	39	40		

Eight Month

Day	Day	Day	Day	Day	Day	Day
1	2	3	4	5	6	7
8	9	10	11	12	13	14
1 Kings 12:32,33 Jeroboam idol worship for political reasons (3235) 15	16	17	18	Baruch sat under an oak tree and wrote 2 epistles. He sent them to the exiles in Babylon (3521) 19	20	21
22	23	24	25	26	27	28
29	30	31	32	33	34	35
36	37	38	39	40		

NINTH MONTH

Day	Day	Day	Day	Day	Day	Day
1	2	3	4th year of king Darius; the people did not listen to prophets. 4	5	6	7
8	9	10	11	12	13	14
15	16	17	18	**Ezra 10:9** Men gathered in Jerusalem on rainy day pertaining to strange wives. (3633) 19 20	21	
22	**Haggai 2:18** Temple foundation to be laid. **Isaiah 34:8** The day of God's vengeance and recompense 23	24	25	26	27	28
29	30	31	32	33	34	35
36	37	38	39	40		

Tenth Month

Day	Day	Day	Day	Day	Day	Day
Gen 8:5 Top of mountain became visible. **Ezra 10:16 Men seperated** from from their alien wives **Ezek 40:1** Future temple specifications 1	2	3	4	**Ezek 33:21** Jerusalem attacked. **Ezek 33:21-33** Struck dumb by God; in the 12th year of captivity. 5	6	7
8	9	**Ezek 24:1-14** In the 9th year, the king of Babylon reached Jerusalem **Zech 8:19** Message from Yahweh to his people. 10	11	12	13	14
15	16	17	18	19	20	21
22	23	24	25	26	27	28
29	30	31	32	33	34	35
36	37	38	39	40		

ELEVENTH MONTH

Day	Day	Day	Day	Day	Day	Day
Gen 8:6,7 Raven sent out **Deut 1:3** Moses spoke to the people for the last time. **Gen 8:6,7** Noah sent out a dove. (1657) 1	2	3	4	Moses died **Deut 34:5** (2655) 5	6	7
Gen 8:8 Noah sent out dove. 8	9	10	11	12	13	14
Gen 8:10,11 Dove return with olive leaf. 15	16	17	18	19	20	21
Gen 8:12 Dove did not return 22	23	24	25	26	27	28
29	30	31	32	33	34	**Deut 34:5** Mourning ends for Moses 35
36	37	38	39	40		

TWELFTH MONTH

Day	Day	Day	Day	Day	Day	Day
1	2	**Ezra 6:15** Second temple completed 3620 3	4	5	6	7
8	9	10	11	12	**Esther 9:1,10** 10 sons of Haman killed by Israelites. 13	**Esther 9:15,16** Israelites killed 75,300 men. 14
15	16	17	18	19	20	21
22	23	24	**Jer 52: 31-34** Jehoichin released from prison by Evil Merodach in the 37th year in exile to Babylon **2 Kings 25:27-30** 25	26	27	28
29	30	31	32	33	34	35
36	37	38	39	40		

CHAPTER EIGHT

NEED TO KNOW – A RECAPITULATION:

As an adult, everyone has an ultimate responsibility for his or her own life, but most of the younger adults today are not interested in the word of God. Their lustful eyes are somewhat transparent to their souls, which is causing them to commit many sinful acts. They are lacking the wisdom, knowledge and spiritual guidance - or they are gravitated towards material things of the world.

PULSE POINTS:

- Mankind has written in their text books that humans have only five senses notably seeing, hearing, smelling, touching and tasting. In selecting those five senses, it may have been a case of lasciviousness, because the writers fell short of listing the other senses of the human race. They are the sense of balancing, coordinating, rhythm, judgment, reasoning, emotions and common sense. This knowledge comes from higher consciousness of spirituality with most of it is written in the scriptures, in the old and new testaments.

- Need to know recapitulation is written, for all age groups to understand a bona fide truth that a living God is totally in control of everything in the entire universe, and other worlds at all times.

- Man and mankind should acknowledge that *Yehshua* was the same person who was on earth and blessed Abraham. Today He remains the high priest after the order of *Melahszte*- meaning peace and righteousness with a position of a highest priesthood in the kingdom of Yahweh. *Yehshua* is empowered and entrusted with the authority to judge the world, and to atone for the people who are foreordained to Him. When *Yehshua* returns to New Jerusalem, He shall have the honor to sit on the throne of righteousness as king of all nations on earth.

- Portentous and Prophetic events will be occurring from the year 2012 through 2014. The primary reason for the extension of time is due to the Gregorian calendar shorting 115 days in the Prophetic year. On the Prophet calendar, it had started in the first month in the year 5509. Christ was crucified on the cross on the day of the Passover in the evening on the 13th day on a new moon. That day was very dramatic, and it is expected to be phenomenal in some part of the world, as the Almighty God see it fit by sending a disaster to that nation. The second cosmic event shall be at the end of a twenty one year Prophetic cycle. (See moon matrix chart). The Cosmic cycle is the last day in twenty one years. A Prophetic year or an earth year comprises of 480 days. A 21 year cycle x 480 days = 10,080 days

- Gregorian calendar conversion is 10,080 ÷ 365 days = 27 years, 6 months and approximately 5 days. Therefore, previous disasters and future phenomena's can be calculated either subtracting or adding 27 years, 6 months, and 5 days in order to pinpoint an event that had taken place, or an event that will be taking place in the future.

- In Genesis 7- 9 Noah was not a boat builder, neither his sons. In that case, God may have assigned angels to work with Noah to the task of building the Ark, gathering up of animals, birds, eggs, young and matured animals to embark unto the Ark.

- Hosea 12:13 states, God had preserved the dead body of His servant Moses, but there was a dispute between Michael the archangel and the devil concerning the body. Jude 6:9. In Deuteronomy 34:6 it states, that Yahweh had Moses buried in a valley in the land of Moab unknown to men. Logically, if Moses grave site had been made known to men, then it would have become a shrine or a biblical sight- seeing place to the public; and maybe man would have evoked his spirit from its resting place.

- *Yehshua* was earthbound for a period of forty days following his resurrection five hundred saints arose from the graves and went into Jerusalem.

- *Yehshua* shall be upon the throne with a counsel of peace being between Himself and Yahweh. Zechariah 6:13. From the time that nations had occupied the land of Canaan, *Yehshua* was designated by God to be the king of *Shalem*.

- Moses, his brother Aaron, along with his two sons was ordained by God to be a priest from the tribe of Levi. They were to fulfill the laws of the priesthood, including the daily offerings and sacrifices in the temple. All animal sacrifices were subsequently abolished when *Yehshua* was sent to earth as a sacrificial lamb with intending to stop the sacrificing of animals. When the second temple was destroyed in the year 4009, it was at that time the daily sacrifices were removed; an abomination was set up in the temple. Daniel 12:11.

- John the Baptist was the son of Zechariah the prophet. *Yehshua* and John the Baptist were coeval - but only months apart. *Yehshua* never grew His hair shoulder length as it is depicted in pictures all over the world in the Christian communities. *1 Corinthians 11:14 states, that according to nature it is a shame for a man to grow long hair.*

- Luke 3:31 states, that the genealogy of Nathan, brother of Solomon, son of King David and grandson of Jesse is thirty five generations after Adam, the son that was created by God. Joseph, the earthly father of *Yehshua* was seventy five generations after Adam and the fortieth generation after Nathan. Do not be confused with the name Jacob, because the first Jacob had a son called Joseph, and the other Jacob also had a son named Joseph, who was the husband of Mary the mother of *Yehshua*.

- God created in the beginning two immortal creatures on the earth and in the sea. The Behemoth conceals himself in the jungle or woods under shady trees. He feeds on grass as the ox does. His strength is in his stomach muscles. His bones are strong as iron and he is constantly drinking plenty of water. Job 40:15-24. Some talk shows may have described that beast as *big - foot*.

- The Leviathan is located in the depth of the underworld and travels the seas. Isaiah 27:1; in Ezekiel 29:3, his description befits a dragon like creature, that spits flames of fire out of his mouth and smoke that comes out of its nostrils. The Leviathan's armored body cannot be penetrated with anything, because of the strong scales, that is covering him are close-up together. He can boil the depth of the sea with his flaming fire, which comes out of his mouth. Job 41:1-33. This sea creature maybe located in the depth of the underworld, and he also travels in the Nile River.

- Man's flesh is made up of worms [maggot] and dust of the earth. Job 7:5; Isaiah 14:11; 66:24; Mark 9:44.

- 2 Samuel 5:4 indicates that King David was thirty years old- thirty nine years and four months on the Gregorian calendar when he began to reign. Solomon's seed was cut off from the Messiah, when he turned to other gods of his enumerable wives that he housed in his kingdom. Solomon subsequently became a mason, and his seed is known today as many Jews, including the Ethiopian Jews who had made a mass exodus to the treasured land; and they are the bloodline of Solomon, 1 Kings 12:16, 19. They are not the seed or the blood line of King David, but some Jews may be from the blood line of King David and not being aware of it.

- The almighty God had pleaded to his people in the scripture, that they should come out of her – meaning that they are trapped in that particular pagan religion. He shall therefore destroy the very last one of those *Edomites* in the tribulation period. Their flesh will be falling off of their bodies in the last days.

- Zechariah 5:1-3 states, that the prophet Zechariah had seen a UFO shaped like a scroll in the sky, so he gave a clear description of its size as twenty cubits by ten cubits.

- Evil persons are very disgruntled from the time they come out of their mother's womb. The spirit of Lucifer is given to them, so as to do the wicked things in the world, same as their father the Devil.

- Yahweh created more than one world; Hebrews 11:3 states that the *worlds* were framed by the word of God.

- Science established that the universe came into existence with a big –bang theory. In 2Peter 3:10, the scriptures states that the heavens shall pass away with a big- bang and the elements shall melt with the heat- a similitude to the beginning of God's creating the worlds.

- Lucifer's seed is described as humankind or hybrid [not humans] and mother earth are infested with these beings that appear to be the normal human race. Their similarities pertaining to the aspects of life on earth are the same, except they have the spirit of Lucifer within them and of the world. When they are expired, their ghosts roam the earth at times until their final day of judgment or the Creator may place their spirit in a body of their kind.

- Man derived from the seed of Adam, and when he dies his soul returns to the Creator. God may use that soul to place it in another body that He chooses on earth.1 Corinthians 15:38.

- The moon turning into blood and the sun darkening depicts a lunar eclipse. During that period of time when it was written in the scriptures, there was no vocabulary of the word eclipse, but only the words of God. The moon will not give light Isaiah 60:19, 20; Ezekiel 32:7, 8.

- New oceans and new rivers shall be located in the deserts .This may be created by a tsunamis caused by great earthquakes along with the large boulders of hail falling into the oceans and seas. Isaiah 34: 4; 13: 13.

- The splitting of Mount Olives in the land of Israel maybe caused by the great earthquake; and so is the flattening of mountains that would be a refashioning of the landscapes and contours.

- All dreams are not from God. A multitude of words spoken during business hours or in time of consciousness may come in a dream in the slumber at night or day. Ecclesiastes 5:3; Proverbs 10:19. Those dreams may not be remembered when one is awaken from their sleep.

- Daniel 12:7, speaks about time = 360; times = 720; half time = 180; Revelation 13:5 speaks about the beast being given power for forty two months, which is equal to 1,260 days, (42x30 days); Daniel 12: 7 states that the man clothed in linen shall disperse the chosen ones for a time = 360 days; times =720 days; half time = 180 days, totaling 1,260 days. Revelation 12:14 states that a woman being given two wings of a great eagle to fly into the wilderness to escape from the beast for a time = 360 days; times = 720 days; half time = 180 days, a total of 1,260 days or three years and five months according to the Lunar calendar. The action of God in foreordaining an earthly decree at times, He uses the date on a nation's calendar- as the lunar calendar in the Babylonian days.

- The Lunar calendar was adapted in biblical times after the death of Cain who died at the age of three hundred and sixty years old (360) according to the lunar calculation. The Prophetic age of Cain's death was 270 years. The conversion is 270 x 480 =129,600 ÷ 360 = 360 years.

- A pure language will be spoken at the beginning of a new created earth. Zephaniah 3:9.

- Jerusalem will have a new meaning to its previous name. "*THE LORD IS THERE*" Ezekiel 48:35.

- Yahweh possesses twenty thousand chariots [20,000 flying objects] in his kingdom. Psalm 68:17, 33. Most of those crafts travel throughout the worlds at a very high propulsion rate of speed- no crafts on planet earth can be compared to a heavenly chariot.

- Let it be known that by law if the wife divorces or separates from her husband she cannot remarry unless the husband is deceased. They shall marry only in the Lord- meaning that they must have the spirit of God within their body and soul. This is defined as an authentic marriage in heaven. A fundamental question is - who is sanctioning your marriage, if your spirit is not from the Almighty God?

- Manna is the food of angels. Psalm 78; 25

- Job 30:30 states that Job skin is black upon him and his bones are burned with heat.

- Galatians 3:10-13 states that those who are under the works of the old law are under the curse- means that if one is continuing to do those things in the old law, they are not justified with those laws in the sight of God. *Yehshua* came and redeemed us from the curse of the law by dying on the cross. Do not be confused with the old laws verses the New Testament; and the mosaic laws with the commandments of Yahweh. *Yehshua* came down to earth to fulfill the laws and not to destroy the prophets or the law. He states that God's commandments will stand forever, but that the old laws were obsolete, and they were replaced with the new covenant. Psalm 19:7, 8. In John 14:15, He states that if you love God, obey His commandments not the old law. *Yehshua* never did mention that one should obey the old laws.1John 5:3. In Matthew 11:30, Jesus made it known that he dinot come to burden the people with the old laws, but to make their live enjoyable and worthwhile living.

- Two thirds of the worlds' population may perish. Ezekiel 5:12, 16, 17; Zechariah 13:8.

- *Yehshua* was a Nazarene and He was under a covenant not to drink wine. However, on His return to earth He shall be drinking wine with His chosen followers in the temple of God. Luke 22:18.

- 1 Timothy 4:4, 5 states that every creature of Yahweh is good and nothing is to be refused if it will be received with thanksgiving, or it is sanctified by the word of God and prayers.

- The river Nile that supplies water to the land of Egypt shall be dried up after the great earthquake, prior recreation of the earth. Jeremiah 51:36; 50:38; Isa 19:5; Isa 44:27.

- Idols, images and witchcrafts will be gone from the Treasured land. Micah 5:12.

- Colossians 2:16, 17. Let no man judge you in meat or drink in respect to a holyday and new moon or Sabbath days, they are a shadow of things to come. Things to be eaten must first be blessed.

- Colossians 2: 18 states that one should not worship angels.

- Deuteronomy 2:5 Yahweh warned His chosen people not to live in mount Seir (Edom), because it is the home of the Edomite (Esau); and He will destroy it at the end of age. Ezekiel 35:2-3; Amos 1:11; Deuteronomy 2:45; Obadiah 1:17 Genesis 36:8.

- Amos 9:8 God promised to destroy the land, but not a total destruction. He will spare a remnant of Jacobs's children Israel and then He shall separate his chosen ones from among the other lands; God will shake the earth (earthquake)like shaking a sieve and not a pebble (chosen ones) will sift through (will die) as He had done in the land of Egypt with his first born.

- 1 Corinthians 6: 9, 10 states that the effeminates, fornicators, idolaters, adulterers, thieves, covetous persons, drunkards, revilers and extortionists will not enter the kingdom of Yahweh.

- John 7:70-71 states that *Yehshua* had identified Judas Iscariot as an *Edomite*, a descendant of Cain and one of His twelve disciples that betrayed Him to the Romans.

- Revelation 20:3 the devil will be cast down into the bottomless pit for one thousand years according to the Prophetic calendar, and 1,115 years on the Gregorian calendar. It is equivalent to 480,000 days and in the Prophetic year 5992 or 2633 on the Gregorian calendar, if it is still in existence.

- Revelation 9:15-18. For a short period of time the Devil will be released for one year and one month (520 days) on the Prophetic calendar; 1 year.4 months on the Lunar calendar of 360 days or 1 year. 4 months on the Gregorian calendar.

- In re-examining few words that are ending with *'ism'*, it has nothing positive or uplifting towards the glorification of the Creator of the heavens and the earth: Fascism; favoritism; confusionism; paganism, federalism; fetishism; Marxism; separatism; Zionism; anti-Semitism; communism; socialism; exorcism; Judaism; criticism; Buddhism; Islamism; capitalism; sudism; vandalism; Hinduism; Roman Catholicism, sadism, imperialism, colonialism and more.

- In the Treasured Land and with a New Jerusalem there shall be no tourism, but a place for prayers in the house of God, from one new moon to another and from one Sabbath to another. Nations shall worship Yahweh in Jerusalem. Isaiah 66:23; Ezekiel 44:9; Zechariah 14:21; Isaiah 35:8; Isaiah 52:1.

- Joseph, Mary and infant Jesus escaped to Egypt from King Herod. When Herod had died twelve years later, they returned to the land of Israel and settles in

Nazareth. That prophesy was fulfilled – out of Egypt I called my son. Hosea 11:1; Matthew 2:15-23.

- Isaiah 20:23 God spoke to Isaiah saying, "*Go loose the sackcloth around your waist and take off your shoes from your feet.*" Isaiah did so, walking naked and barefoot three years in Jerusalem.

- Yahweh put words in the mouth of prophets and others including a donkey. 1 King 10: 7-9 Queen Sheba of Ethiopia visited Solomon in Israel, and God put these words in her mouth. "*Blessed are your servants who are always present before you to hear your wisdom. Blessed is Yahweh your God who loves you and put you on the throne of Israel, because God has loved Israel forever, and He made you sovereign to do right-ruling and righteousness.*"

- Exodus 4:10-12 God said to Moses, "*Who made the mouth of man, the dumb, deaf and the seeing or blind? Have not I the lord? Now go and I will be with your mouth and teach you what to say.*" *Isaiah 50:4;*

- Numbers 22:28 God opened up the mouth of an ass and she said unto Balaam "*What have I done to you, that you strike me three times?*"

- Some Israelites and Jews are converts, while others do not know their true roots. The Canaanites are the heathens that served idols; foreign Gods and images. To reiterate, they are the seed of Lucifer and his first born son Cain, who killed his brother Abel, the first born son of Adam.

- *Matthew 2: 9, 10. An object that appeared in the sky and witnessed by the three wise men, were actually one of many heavenly chariot with a very bright light, that guided them to their destination to see the new born King.*

- John 8:21; 7: 34; 13:33.*Yehshua* said that, wherever He goes no man can follow. He was referring to His resurrection and His returning to heaven.

- When you die your body goes back to the dust awaiting judgment day, and the soul either goes to *Sheol* or shall be returning to the creator to be placed into a body of His choice.

- Revelation 4:5; 1:4, at the end of the age Yahweh shall destroy the worlds, the heavens and recreate new ones including earth. Those extra beings are aware that their time are drawing near, and they are seeking new place to dwell in - same as mankind on earth that are venturing out into the vast universe seeking a planet that has traces of life on it including water.

- *Nicolaus Copernicus, 1473-1543BC, and Galileo Galilei, 1564-1642, both astronomers said that the earth is in motion around the sun. Their theory carried credence to convince others to accept their claim. Scriptures gave various statements that the sun raises and the sun goes down. Some Religious and social paradigms in the fifteenth and seventeenth centuries objected to this fact.*

- Joshua commanded the sun and the moon to stand still in the midst of heaven during the battle with the Amorites. Joshua 10:12-14. The earth is known to be rotating on its axis. The scriptures confirm that this is an accurate statement, and it is a consistent assessment that the sun rotates around the heavens.

- Having dreams about vicious dogs; snakes; cockroaches and rats are warnings that one should be aware of that they have enemies.

- The entire house of Israel and those who are in the grave will be brought back to life; and they shall be taken to the land of Israel. Ezekiel 37:11-14; Hosea 13:14.

- Elijah the prophet had already returned before the second temple destruction and before the tribulation had started. That took place immediately after the temple was desecrated; abomination committed and the exile of the children of Israel in the year 4009. Matthew 17: 10; 11: 4. *Yehshua* stated that Elijah had already come back to earth; and that no one had recognized Him. Matthew 17:12; Malachi 4:5; Mark 9:13. Elijah was John the Baptist reincarnated.

- At the end of the age men shall choose, and they shall have the desire to die rather than bear the tribulations, but there will be no death. Revelation 9:6; Jeremiah 8:3.

- In the latter days evil men will attempt to escape from severe tribulations that shall be taking place, but Yahweh shall find them and kill them all Amos 9:1-5.

- The old Jerusalem will be destroyed. The land will turn into burning pitch, the dust into brimstone; and it shall be wasted forever. Isaiah 34:8-1. This explains that the crude oil below the earth will be surfacing to fuel the burning lake of everlasting fire, same as in the days of Noah's flood, when the fountain of the depth of the earth released its waters to flood the earth in the year 1656, a Prophetic date.

- *Yehshua* was thirty years old when he was crucified on the stake; and He was foreordained for one generation of thirty years without a seed to follow.

- A church may have a mixed congregation of good and evil-doers; Judas Iscariot one of the twelve disciples had the spirit of Lucifer controlling him; and he subsequently betrayed Jesus that led to His crucifixion on the cross.

- The tribulation periods commenced when Titus and his armed forces desecrated; and committed abomination in the temple of Yahweh in the year 4009, spoken by Daniel.

- The Greek word *Xristos* (Christos), in English (Christ) means the *"Anointed one."* His Hebrew name is *Yehshua*.

- The Messiah shall be revealed when the world will be preoccupied with mundane things such as: in the work place; in the field; at school; in a bar; at church; at a mosque; in a synogue; at the theatre; at sea; at the Library; on a flight in the air, and occupied with other activities. Luke 17:30.

- Terrible plagues shall befall all the people who accept the mark of the beast and those that blasphemed the name of Yahweh. The mark of the beast may come in any type of form - taking a vaccination on the arm, or a given seal in a tattoo form at birth.

- Revelation 14:17. The number of the Beast is six hundred, three score and six (666), Revelation 13: 18. That number depicts a man according to the calculations of his age on the Prophetic and Lunar calendars.

- In Genesis 9: 26, 27, it states that Canaan and Japheth, along with the son of Noah shall be servants of Shem. And Japheth shall live in the tents (synagogues) of

Shem. Let him that has understanding count the number of the Beast .Revelation 13: 18. The hidden truth secretly lies within a spiritual vault. So then, to adept that combination, the tumbler must be rolled forward at 360 degrees in which things are to come; and then it must be turned counter- clockwise at 480 degrees, same as the earth's rotation, so as to decipher that code. *In a matter of caution, anyone that chooses to accept the mark of the beast of any type, engaging in any purchasing or business transactions will be damned.*

- *If your spirit is from the Almighty God, so then, do not be deceived in taking any sort of vaccination that will be permanently remaining in your blood stream; and maybe altering your DNA, so that you may become affiliated with the Devil's schemes.*

- There will be jaw dropping in awe to the disclosure of the man who is known as the beast which is mentioned in Revelation. This mysterious man will be severely wounded in battle, but shall be fully recovered, so as to fight the final battle between good and evil.

- Shem's death was at the age of 500 years old.
  Prophetic days is 500 × 480 = 240,000 days
  The number of Shem is 240,000 ÷ 360 = 666.
  On earth, he is being worshipped as Ha Shem- meaning the name.

- *The ages of the below mentioned biblical characters were recorded on the Prophetic calendar of 480 days in a year. The conversion to the Gregorian calendar is as follows upon their death:*
  King David was 70 years old (92 years old)
  King Solomon was 52 years old (68 years old
  Moses was 120 years old (157 years old)
  Aaron was 123 years old (161 years, 7 months)
  Meriam was 93 years old (122 years old)
  Enoch was 365 years old (480 years old)

- Heathens are Pagans, and Gentiles are some of Yahweh's people who do not believe or accept *Yehshua* as the son of God. They are called the anti- Christ. Anti Christ is not only an individual man or person, but there are many others, that are not believers in Christ and His resurrection from the grave. Heathens were the nations who hated the Gentiles. 1 Corinthians 12: 1, 2 articulated that the Gentiles were carried away unto dumb idols, even as they were taken away by the Romans into exile. Gentiles are the seed of Abraham and his heirs according to the promise. Galatians 3:29; Genesis 12: 3; Ephesians 3:6.

- The seed of Jacob are the tribes of Israel, and they are Yahweh's chosen people consisting of one hundred and forty four thousand. They are the ones who shall be dwelling with the Messiah on His return at the end of the age. The chosen ones will be close to him at all times in the new world to come for a thousand years.

  However, at the end of tribulation period Yahweh shall cover the heaven, make the stars not shine, blanket out the sun causing it not to give its light; also make all the bright lights of the heaven dark over earth, Ezekiel 32: 7, 8.

- Genesis 4:15......... and the lord did set a mark upon Cain. That mark could have been a curved nose. Throughout the years the Canaanites have intermingled and had inter- marriages. The angels of the lord shall identify them from among those who are living and those who are sleeping in the dust, so as to face the final judgment.

- There are over a hundred nations and islands that are not notable to be ravaged by deadliest natural disasters, as earthquakes, famine, tsunami, contractible diseases, floods, landslides; or other communicable diseases. (See website, countries of the world)

- When the heaven opens up the angels of God shall ascend and descend upon the son of man. Some will be taken- meaning, that they will die in that process and their remains shall be left on earth.

- Hebrew means crossed over; Abram crossed over from the Euphrates River to Canaan.

- The word Christian was first used in Antioch. Prior to that, they were called Disciples; Apostles; brethren or the people of the way; Apostle's means sent out.

- The Eagle is a symbol of Rome and which is similar to the eagle symbol at the White House.

- 1 Corinthians 14: 33-35, clearly articulated that women are not permitted to speak in the church.

- Everything was set in place by God at the beginning of His creation and subsequently awaiting His command to be executed. Examples are the Rods of Moses and Aaron.

- Isaiah 38:8; 2 Kings 20:9-11 – God turned backwards the sundial clock or King Ahaz to ten degrees. In Jan 11, 2011 it is said that the magnetic pole shifted ten degrees.

- Hosea 7:12 – States that Yahweh will spread His net and bring down birds of the heaven- a similitude used to describe His action against Ephraim. Hosea 5:5

- Only the chosen ones know the new song of Jerusalem. They shall be singing that song when the joining with the Messiah, the Anointed One in the New Jerusalem. Psalm 96.

- God created man to live and not to die. Man has sinned causing his eternal life to be relinquished until the return of Yehshua. However, each and everyone are solely responsible for his actions, his soul; and he shall be judged accordingly on the day of the final judgment.

- *Generally, it would be effortless or it may be normal to some people to commit most of the following sins: adultery; robbery; crookedness; curses; deceitfulness; harmful; negativity; always with excuses; vengeful; witchcraft; heresy; greediness; glamorous; insincere; criticize, but has no solutions to offer; plot evil; rootless. Betrayal; insincere; malicious; malice; boastful; materialistic; borrows, and not repays; an evil eye; unscrupulous; destructive; slyness; slanderous; lover of drugs and alcohol; lover of pornography; vulgarity; jealousy; gossiper; backstabber; corrupt; disruptive; vindictive; uncooperative; rape; selfish; false witness; proud, and arrogant. Injustice; enslave others; creates strife; blame other for their lasciviousness; uses and discards each other. Gluttonous; swearing by the heavens and earth; threatening one another; holding grudges; liars; carries on filthy conversations; abominable idolaters; creating ungodly laws; inventors of evil things; rumor mongers; ungratefulness; reprobate mind; pernicious ways; lustful; blasphemers; disloyal; haughty; swears; dislikes another for no reason; spitefulness; lawlessness; self – assumers; breaking and not building; unthankful; fierce and quarrelsome. Envious; unmerciful; unforgivable; harassment; hypocritical; laziness; covetousness; discontentment; troublemaker; dangerous; bribes; sponges on others; consorts with, man to animal, man to man, and women to women; mischievous; murder; hatemonger; a spirit of vexation; always unpleasant; rejoices over one's downfall; bear false witness; sadism; dishonorable; violent, hasty and always bitter. Uncharitable; a sole lover of money; cruel and wicked, fornication; reckless; theft; ruthless; sorcery; wrath and anger; frivolous moods; whoremonger; unfair; disrespectful to one another; ill-mannered; brawlers; disobedient to parents; unholy; without natural affection; trucebreaker; traitor; lovers of pleasures; extortion; defrauds; abusers; effeminate; revilers; drunkards; practicing Satanism; seeking spiritual answers on a Ouija Board; ungodly; rebellious to the truth; perverted lips; false balance and mischievousness.*

- All the above cherry- picked iniquities are self-destructive to the soul, body; and also they are very destructive to others, including those who are close to them.

- It is a great struggle within our being not to speak the truth when confronted. Evil always knocks on your door, but it's up to the individual to rebuke it. Evil will take control of your mind and body if you allow it to happen. The leaders of nations who are believers in the Almighty God have to pray for the good of their nation, that is, if that nation is on the wrong path and needed to be redeemed.

- Display a positive attitude, because evil actions will have a setback. No evil goes unnoticed by the Creator, because all living soul is being watched. All iniquities are recorded in the soul and they will be revealed on judgment day, even though one may not remember all their wrong doings.

- Suffer not your mouth to cause you to commit a sin. Two persons together are better than one, because if one ever falls, the other may be able to pick them up. Love is not delighted in evil, self-seeking or destructiveness, but it comes from the creator of the heavens and the earth. Love your neighbor as you would love yourself. It is better not to make promises and not follow through; it is difficult to say, "I am sorry;" The question is, what positive contribution have you made toward society and humanity?

A commitment to honor man's time and deadline causes: Stress; accidents; hypertension; an anxiety attack; and other serious ailments. The Prophetic calendar brings about: Joy; comfort; satisfaction; and spiritual cleansing to one's body and soul.

Every living organism has a purpose in life on this earth. Every living thing including plants, animals and the stars worship its creator in their own way.

- It has been recognized that persons with less stress or a righteous person in the Lord and not being abusive to their body ages very gracefully.

- Should one be dishonored and not recognized as an author, or be degraded or disparaged because they are not a scholar; theologian or scientist? Yehshua was a carpenter by trade before He revealed His true identity. He did not attend an academy of biblical learning, but He taught some scholars in a synagogue to their amazement at the age of twelve years old when He had returned from Egypt.

- Revelation 7:5 indicated that Yahweh has chosen one hundred and forty four thousand of the tribes of Israel. The chosen tribes are: Judah; Ruben; Gad; Asher; Naphtali; Manasseh; Simeon; Levi; Issachar; Zebulon; Joseph and Benjamin. Each tribe shall be consisting of twelve thousand. The tribes of Ephraim and Dan are not mentioned or allocated with equal amount of followers as the other tribes. Revelation 7:13-15 states that God will merge the tribe of Ephraim with the house of Joseph, and they shall be as though Yahweh had cast them off and this will strengthen the tribe of Judah. Zechariah10:6; Psalm 78:67, 68.

- Hosea 5:9 indicates that the tribe of Dan will not be of twelve thousand, but will be allocated to the east and west sides at the border of Damascus and northward to the coast of Hamath. The gate of Dan shall be located at the east end. Ezekiel 48:32.-1

- *Job 17:12. The world had changed the night into day. The daylight is short because of darkness. On planet earth, they have altered the time the sun rising and when it is setting, so that the Gregorian calendar and the lunar calendar may be able to work in conjunction with their timing, but it is not possible.*

- Luke 12:51 Jesus stated that one of his primary mission was to divide up the people on earth, so as not to bring about peace. Today it is evident that the three main religions that are institutionalized are having their own written laws and many books to substantiate what they are preaching to their people.

- A family may have in their household a mixture of good and evil. Biblical examples are the two brothers, who were conceived by Eve, were Cain and Abel. Abel was righteous, but Cain was evil as his father Lucifer. Also Isaac's two sons, Esau and Jacob were conceived by Rebecca, and they were different in personality. Esau was not interested in God's commandments, but in evil things, while Jacob was more diligent and studious in the word of God. The population in the world has increased to billions of evil and good people.

- *A Simple Tablet of Yahweh's Calendar – 480 Days Calendar:*

One day = *24 hours approximately 12 hour day; 12 hour night depending on location.*

Six days = *Work*

A seventh day = *Rest (completion)*

Forty days = *One month*

480 days = *One year*

Six years (2,880 days) = *Work the land*

A seventh year (3,360 days) = *Rest the land*

Forty nine years (23,520 days) = *Seven sabbatical years*

A fiftieth year (24,000 days) = *One Jubilee year*

The circumference of the earth is 960 mph x 24 hours = 23,040 miles.
2 Peter 3:8; Psalm 90:4
1,000 years to God is one day to human and humankind.
1,000 Prophetic years are 1,115 Gregorian years
A new creation = 140 Jubilee years x 50 = 7,000 years on earth and 7,000,000 million years in heaven.

The earth was in existence, but continues to be destroyed and recreated by God. So then, in every 6999 years it is the approaching Sabbatical years of the universe, and the 7,000th year is its Jubilee year. On earth, every forty nine years is the sabbatical year and the fiftieth is the Jubilee year. The holy land is six years there is sowing and reaping. In the seventh year it was to be rested as per the old law, but it is not a requirement in the New Testament and in the world today.

Humankind had worked six days and had to rest in the seventh day. Those were the laws of Yahweh. He has designed the earth to have one hundred and forty Jubilee years (140 x 50 = 7,000) the same concept and laws had applied to mankind, the land with the other worlds.

- *John 8:44 Yehshua gave a clear and concise description of Lucifer the father of mankind- not man. He said that they are of their father the devil, and the lusts of their father they will do. Lucifer was a murderer from the beginning.*

- The unconscionable evil of mankind demonstrates and identifies the seed of Lucifer that dwells among us today. They are identical to humans, very intelligent and they are having high analytical qualities. They are occupying top positions in Governments, controlling a nation's political arena, Media, financial institutions, Entertainment industries, Real Estate and economic systems and more.

- *Example #1: Prophetic calendar*
  40 days/ 40 night × 24 hours = *960 hours*
  960 hour × 12 months = *11,520 hours*
  11,520 hours ÷ 480 days (one year) = *24 synchronized hours.*
  John 11:9 Yehshua said; are there not twelve hours in the day?

- *Example #2: Lunar calendar*
  30 days/ 30 night × 24 hours = *720 hours*
  720 × 12 months = *8,640 hours*
  8,640 hours ÷ 360 days = *24 hours*
  A twelve- month moon cycle amount to 340 moons as indicated on the moon chart.

- A question to the scientific community. How can the lunar calendars carry 354 days or 360 days in a calendar year, when in some months of the year there are two new moons in that month, but it is still considered a single calendar month?

THE BIBLICAL FESTIVAL CELEBRATIONS OF YAHWEH WERE:

Passover *(One day)*	*14th of the first month, at evening of* *the 13th (Prophetic calendar) Lev23:5; Deut 16:1*
Unleavened Bread *(Seven days)*	*Starts at evening of the 14th day of the first month, and ends on the 21st day of the first month at sundown. Ex 13:3; Deut 16:3; Lev 23:6*
Blowing of Trumpet *First of all fruits*	*Starts at evening of the 40th day of the 6th month, which is the seventh Month, on the first day. Num 29:1; Deut 26:2; Ex 22:29; Lev 27:26; 23:24*
Day of Atonement: *Trumpet blasts and total* *Fasting, in a Jubilee year*	*Starts at evening of the 9th day of the seventh month; which is the beginning of the 10th day. Lev 16:29 Every 50th year,*
Feast of Tabernacles *(Booths)* *(Seven days)*	*Evening of the 14th day, which is the beginning of the 15th day, and lasts seven days. Ezra 3:4; John 7:2*

Countdown to the Next Biblical Event
(Prophetic calendar)

The old laws of the testament are as follows:

1. *One day after the unleavened bread, on the 21st day of the first month, the priest waves a Sheaf on the Seventh Sabbatical day (49 days). On the 50th day it is Pentecost. Meat offering is on the 24th day of the second month, in the evening of the 23rd. Lev 23: 15, 16*
2. *One day after Passover to the Day of Trumpets, on the first day of the 7th month are a total of 227 days, (In the evening of the 40th day of the 6th month) Num 29:1*
3. *One day after the trumpets to the Day of Atonement, on the 10th day of the 7th month is, 9 days, in the evening on the 3rd day of the week. Lev 16:29, 30*
4. *One day after atonement, to the days of tabernacles when families slept outdoors is, 7 days. It begins on the 15th day of the 7th month, at evening of the 14th day Lev 23:34, 35, 39.*

These were the mandatory festivals required to be observed by the children of Israel: the tribes of Judah, Levi, Simeon, Asher, Gad, Dan, Joseph, Manasseh, Ephraim, Reuben, Zebulun, Issachar, Naphtali, and Benjamin. However, the tribes of Israel are scattered around the globe, and animal sacrifices are no longer required from the time of the death of Yehshua. Psalms 51:16; Daniel 12:11

CHAPTER NINE

UNANSWERED QUESTIONS

This chapter of unanswered questions by many people is geared to reveal the hidden truth about why the Creator is taking certain decisive actions, when His commandments are not adhered to. However, the Final Day of Judgment will be at the end of the age, and those who are to live shall stay alive, and the sinners shall be thrown into the lake to burn in the everlasting fire. It also explains why the old law or covenant was replaced with the new covenant. The answers to these questions below are found in scripture verses in the old and the new testaments:

Q. Why 40 days in a month in a Prophetic calendar year of 480 days?

A. *The solar days in a year is   480 days*

    The lunar year is -------- 360 days

    Earth days in a year is -- 480 days

      Total   = 1320 days

The Prophetic conjunctional days are 1320 days in total. Psalm 90:4 states that 1000 years in our sight is like yesterday when it is past; and as a watch in the night.

    The sun at 480 days ÷ 40 days   =   12 Prophetic months

    The moon at 360 days ÷ 40 days   =   9 Prophetic months

    Earth at 480 days ÷ 40 days   =   12 Prophetic months

      Total   =   33 Prophetic months

Therefore, 1320 Prophetic days ÷ 33 = 40 Prophetic days or one month that is working in conjunction with the sun, moon, planet earth, the two seasons of summer and winter.

Q. Why is the earth suspended in space, and should not be moving off its axis?

A. Because heaven is the throne of God, and the earth is considered to be his footstool. Then, a footstool is affixed to the heaven by a plum-line from the seventh heaven to the earth. Isaiah 66:1; Matthew 5:34, 35. The meaning of the word stool (footstool) is used for urinating or defecating of matters. In Genesis 3:17, God cursed the earth.

Let it be known that the sun and the moon gravitational forces, along with the heavenly vertical plumb- line is compassing through Jerusalem, which is considered to be at the center of the earth, stabilizing and suspending the earth into space. The sun and the moon are responsible for the rotation of planet earth, and in maintaining it on its axis. Genesis 28:12, 16

Universal time table:
Prophetic days in a year = 480
Prophetic days in a month = 40
Prophetic hours in a year =11,520 [480 × 24]
Solar days in a year = 480
Solar days in a month = 40
Solar hours in a year =11,520 (480 ×24)
Lunar days in a year = 360 [30 × 12]
Lunar days in a month = 30
Lunar hours in a year = 8,640 [360 × 24]
Psalm 139:12 states that the darkness cannot be hidden from God, because the night is like day to Him.

Colossians 1:26; 1Corinthians 2:7 concludes that even some mysteries which have not been disclosed from ages, and to previous generations are now made manifest to God's people.

Q. Why did Adam and Eve hid in the Garden of Eden when God had called out to them?
A. Because for the first time two selves- taught, self- sufficient and independent human beings, who were given a commandment not to eat from the fruit of the tree that was in the middle of the garden, had disobeyed God's commandment. *Genesis 3: 16. The tree of righteousness represented God. The tree of good and evil represented Lucifer the serpent.*

God then told Lucifer that his seed (Generation) and Eve's seed shall be enemies forever. Cain who was from the Devil's seed, killed his righteous brother Abel who was the first son of Adam, but Cain was not his son.1John 3:12.

God told Abraham not to mix his seed with the people from the land of Canaan, because they were the seed of Lucifer. *Deuteronomy 7:1; John 8:44. "You are of your father the devil.* In today's world, the seed of Lucifer is the ultimate problem of society recognized as humankind or mankind on earth, and beneath the earth.

Q. Why was the Golden calf made in the absence of Moses?

A. In the wilderness, the people of Egypt adapted the lunar calendar which consisted of 360 days. Moses calendar was the Prophetic calendar comprising of 480 days. The lunar month carries 30 days, and the Prophetic month has 40 days.

Moses may have told the people that he will be away for a month, but the multitude of people of Egypt, along with the children of Israel used the lunar months of 30 days. This may have been interpreted by the people in the wilderness as a month. Moses subsequently returned from the mountain ten days later, and was confronted with vile behavior by the multitude of people in the wilderness.

Q. Why some scientists believe in evolution and righteous men do not?

A. Apes have many species as every living thing does, but they are not extinct as far as the world knows. Man did not derived from the family of apes, but man was created by God from the dust of the earth and he has a soul, but animals do not. However, the majority of humans believe in creationism, while only a few believes in evolution. Genesis 2:7.

Q. Why twelve months in a year?

A. Monthly shifts were performed by captains and their marshals, that entered service left month by month throughout the year which consisted of twelve months. 1Chronicle 27:2.

Q. In Zechariah 3:1-10 why was Joshua brought before Yahweh in filthy clothes, while Satan was standing at his right side to resist him, when Yahweh was rebuking Satan?

A. In the Exodus, Moses was designated by God to lead the people out of Egypt to the Promised Land. However, Moses fell short of his mission and died in the wilderness, before his mission was accomplished. Joshua, Moses second in command took on the challenge, and he accomplished that task by leading the children of Israel across the Jordan River- with the highest degree of success. God was very pleased with Joshua's performance, so he had promoted him to a higher priesthood for the future temple of God. In the new temple in New Jerusalem, it will no longer require sacrificial animal slaughtering. Joshua was cleansed from all sins, and he was given new garments with a turban on his head for a future role as a high priest, while Satan stood at his right side objecting.

In future, the Messiah shall be the prince of the entire earth with a one world order and one pure language spoken, as was intended at the beginning of creation. However, He shall be a priest forever after the order of *Melahszte*. Psalm 110:4.

Q. At the end of the age, why will the sun no longer be the light of day with its brightness, and the moon will not be withdrawn? Joel 3:15; Matthew 24:28.

A. God shall be that eternal light unto the world, Isaiah 60:19. The sun will not shine or will it be used as an instrument for a calendar, but the moon shall be resumed its celebration at every new moon when God returns to the earth.

Q. Why the Ark of the Covenant will no longer be used in the new world to come?

A. Because the Ark was taken away and hidden in an undisclosed location after the destruction of the first temple. In time to come, God's presence will be felt on earth.

Q. Why did the prophet Jeremiah curse the day that he was born?

A. Jeremiah was a righteous man, and he had witnessed in his lifetime that the children of God were not obeying the laws, the statutes, the ordinances and His commandments. In the outcome he saw the results of sufferings, death, and he became part of it. Jeremiah 20: 14-18.

Q. In God's new creation of the earth, why will there be one pure language instead of many languages on earth?

A. The surviving nations on earth at the end of time, shall announce the name of God correctly, and not call him by many names as it is happening today. Also, the word of God shall not be translated incorrectly or be misinterpreted.

Q. Why Zechariah described a flying object as a scroll (roll)?

A. In the days of Zechariah there were no other words in the vocabulary to describe an un-identified flying object (UFO); and there are twenty thousand chariots in the kingdom of heaven. Psalm 68: 17, 33. Those chariots will be engaging in the final battle between good and evil. Presently, the heavenly chariots pay visits to planet earth and other worlds.

Q. Why Yahweh says in Malachi 1:2 that he hates Esau, but he loved Jacob?

A. Esau in his days had done everything contrary to the word of God.

Deliberately he married the daughters of Canaan; and to this day his seed is attempting to take back the birthright that Esau had given to Jacob.

Q. Why the chief officer in Pharaoh's palace gave new names to Daniel, Hananiah, Mishael and Azariah?

A. Genesis 41, the Pharaoh gave Joseph *Zaphnathpaneah* a new Egyptian name.

It was customary in Egypt when someone was given a high ranking position, that new names would correspond with that high rank attained.

Daniel, he gave the name *Belteshazzar*; Hananiah he gave *Shadrach*; Mishael to *Meshach and* Azariah he gave *Abed-Nego*.

The king of Egypt made Eliakim son of Josiah king, and he changed his name to *Jehoiakim*. 2 Chr 36:4; Nebuchadnezzar changed the name Mathaniah to *Zedekiah* and made him king of Judah.

Nebuchadnezzar may have adapted the custom of the Egyptians.

Q. Why is it against the law of God for one to consume blood?

A. Because it was the blood that makes atonement for the soul; and the life of the body is in the blood.

Q. Why the land of Israel should not be bartered or sold?

A. God stated in Leviticus 5:23 that the land shall not be sold ever, because the land is His own, and that all occupiers are strangers that are only passing through. The laws of the land were not adhered to by keeping the Sabbath of the land after every six years; and the seventh was its Jubilee. Leviticus 25:3. When the laws were broken, the children of Israel were then scattered around the globe, awaiting the return of the Messiah and a New Jerusalem. John 8:39; Romans 9:6; and Galatians 6:16 imply that all the people in Israel are not Jacob's seed.

Proverbs 22:28. Today Sabbath observances are not necessary according to the new covenant, but it will be in the new world to come, and it shall be reinstated.

Q. Why Yahweh told Cain that sin is waiting at his door? Genesis 4:7

A. Yahweh knows everything that men do and plot in their hearts. He knew that Cain had premeditated to kill his brother Abel; and God did witness the slaying of Abel, even though He had enquired from Cain about the whereabouts of Abel.

Q. Why did Abram and Sarai had no children when they became husband and wife?

A. The word of God was established at the beginning of times. Leviticus 20:17, states that he shall bear his sins. God changed both their names from Abram to *Abraham* and from Sarai to *Sarah*. Their sins were forgiven, and in their old age Sarah conceived and brought forth Isaac. They did leave Haran away from their family and out of sight from their people according to the word of God.

Q. In Genesis 3:22, why did God say that man has become like one of us knowing good and evil?

A. Satan was ordained by God, and he dwells in one of the Heavens among the hosts of angels. He has great knowledge of the word of God and the activities in heaven, because he was the only *anointed angel*. Satan has unlimited access to the heavens, and occasionally God uses him to perform as a rod of correction in specific tasks, as in the case of Job that was destructive to his family , but it had a very good ending. At the end of the age, God will no longer require the services of the Devil.

Q. Why the eleven brothers of Joseph wanted him dead?

A. Joseph's dreams were no ordinary dreams, but they were divine disclosures, that he was predestined to become the leader of the family. Reuben was the eldest in the family, but that status was relinquished when he slept with Bilhah, Rachel's maid.

The dreams of Joseph made the brothers suspicious, angry and they became jealous, because their father Jacob had treated them indifferently from their brother Joseph. They recognized that their father's behavior towards them as a threat to the disinheritance of Reuben's birthright. Simeon and Levi were not qualified as leaders of the tribal family, because of what they had done in *Shechem*. They slaughtered the men in the land and Jacob was angry with them. Joseph became the father of two tribes over his brothers, and he inherited double portion of the land.

Q. Why God says to trust in him with all your heart and not rely upon your own understanding?

A. Because He founded the earth with wisdom. He estimated the heavens with an understanding that no human can comprehend through scientific means; or by measurement of time and space. *"Let your trust remains with Him"*.

Q. Why did Job curse the day he was conceived?

A. Job said, *"let the day perish wherein I was born.........Why did I not die from the womb? Why did I not give up the ghost, when I came out of the belly?"* Job 3:1-3,
Satan inflicted sore-boils on Job from head to toes, and he killed Job sons and daughters by staging separate fatalities. However, Job was blessed by Yahweh with seven more children and with lots of wealth.

Q. Why should the teaching of Yehshua, two thousand years ago has been chosen to Echo through the church?

A. Yehshua was a teacher of faith in which He had demonstrated to His disciples on many instances as they walked with Him in the land. So his teachings should be about what He taught, and not about an individual's opinion

God has given the complete revelation to Yehshua to pass on to all, so that men should know Him and draw to Him in faith, which is a free gift from God The faith He revealed to us intended to unite individuals into one belief and as a family in faith.

Q. Why was King David not delegated by God to build His house?

A. Even though David was loved by God, he had committed some grave sins in the eyes of God. He committed adultery with his captain's wife. He arranged to have Uriah killed, and he allowed himself to be deceived by the devil in taking a census, by doing it tribe by tribe without obtaining permission from Yahweh.

Q. Why did Yehshua say in Matthew 16:28, that some of the disciples who were with him will not die until they witness the son of man approaching in his kingdom?

A. Kingdom in that sense means His glorious form and not the old flesh that died. Some of His disciples during that generation did not pass away until after Yehshua had risen from the dead. The disciples had witnessed Yehshua being around in that forty day period of time, before He had ascended to the seventh Heaven. Peter and Judas died before his resurrection, but the others did not die, who had witness his new form of kingdom as He had once told them that He will be raise in three days. Many righteous ones who were in Paradise, including Lazarus arose from the grave, came back to life and went into Jerusalem, before later ascending into heaven. Isaiah 26:19; 1 Corinthians 15:6.

Q. Why was Cain's name not listed in the genealogy of Adam to Jacob?

A. Because Cain was not the son of Adam, but he was the son of Lucifer the serpent of deceitfulness. His genealogy started *from Cain - Enoch - Irad – Mehujael - Methusael -Lamech - Jabal and to Tubalcain who was killed accidentally. John 9:44*

Q. Why were women condemned to death by stoning when they were caught committing adultery?

A. Both the old Law and the new covenant apply to a male and female who was caught committing adultery. Scriptures have no records cited that a man was stoned to death for breaking that law. However, it appeared in the scriptures that the law only applies to a woman, because Eve was the first woman to commit adultery in the Garden of Eden. The law of Yahweh pertaining to adultery still applies to both male and female married couple; and it does not discriminate. When a spouse commits an act of adultery; stoning someone to death is not the law of the land. The accuser is either pardoned or divorced from each other, or it can be a separation.

Q. Why should man trust in the Lord?

A. The creator of the universe can be relied upon to attain results regardless what the task is. God is extremely reliable to accomplish any prayer requested of Him. Joshua and Caleb trusted in God concerning spying out the land of Canaan, but the other ten spies did not. They discouraged the people on their return from spying out the Promised Land, but Joshua of the tribe of Levi, and Caleb of the tribe of Judah, inspired the people in giving them hope for success.

Q. Why was Moses not allowed to crossover to the promised land of Canaan?

A. Moses killed an Egyptian in the land of Egypt. He disobeyed God and struck an important rock in disgust to have water flowing from it for the people who were complaining to him.

Q. Why Jeremiah claimed that King Zedekiah would be carried captive to Babylon, while Ezekiel prophesied that he would not be seeing Babylon?

A. Both prophets were on point with what they had prophesied. Zedekiah was taken captive to Babylon, but he was blinded by king Nebuchadnezzar, so that he was not able to see Babylon.

Q. In John 20:14, and 15, why did Mary Magdalene not recognized Jesus in the sepulcher, and thought that He was the gardener at the burial site?

A. Jesus died and was buried in a natural body, but was raised in a spiritual body. 1 Corinthians 15:44. Jesus adapted various forms in the presence of his disciples and both Mary's.

In 1 Corinthians 15:38 states, that God gives the spirit a new body that pleases Him; and to every seed his own body. Yehshua died in weakness, but was raised from the dead in power. 1 Corinthians 15:43.

Q. In Genesis 9:24, Why Noah say that he knew what his younger son Ham had done to him?

A. When Noah had become sober from his drunken state he was informed, that Ham had slept with his young mother – Noah's wife. Genesis 9:21-23

In Leviticus 18:7 states that the nakedness of your father or mother shall you not uncover, because she is your mother. Leviticus 18:8 states that the nakedness of your father's wife you shall not uncover (undress) because it is your father's nakedness.

Japheth and his brother Shem took a garment; they walked backwards and covered their naked stepmother. Genesis 9:23.

Q. Why is it so difficult for some preachers to reveal the true Gospel to their congregation, knowing that the end is drawing near?

A. Most religions, culture, nations and generations are steadfast in their traditional ways in teaching or upbringings where the gospel is concerned; and it is difficult for them to deviate from that old school. When Yehshua was on earth in the years 3949- 3979, it was stated that He did not come to destroy the law or the Prophets, but to fulfill them. Matthew 5:17.

Yehshua, Isaiah and Hosea used many metaphors in their messages to the people during their time on earth, but all the messages delivered were the same- changing from evil to doing good in order to have a long sustainable life on earth; and being a good leader to others in the world.

It is not the tone of voice of the preacher that is delivering a message, but the text and truth of the message, which should be subsequently analyzed and checked out by those who are listening. The message delivered should not be with the opinion of the speaker or by his own interpretation of the scripture verse, but it must be compatible with scriptures, and not for a preacher's gain or amusement of his audience.

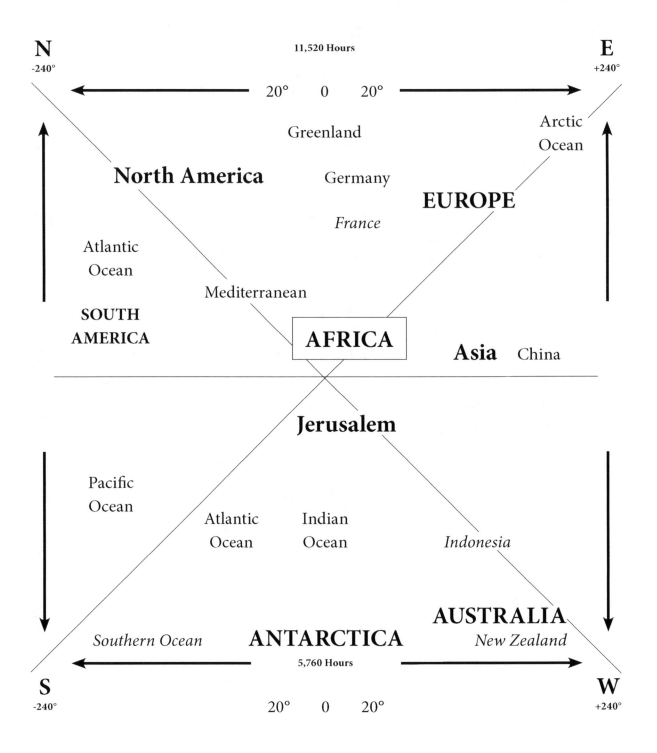

REMAPPING THE WORLD

A simple synopsis of the sun, moon and the earth:

The sun's daily gyration around the earth is twenty four hours (12 x 2 in the northern and southern hemispheres). The earth covers 40 miles every hour during its rotation (960 mph ÷24hours = 40 miles)

The circumference of the earth is 960 mph x 24 hours = 23,040 miles.

The earth's rotating hours in one year is 480 days x 24hours = 11,520 hours.

The total amount of months in a year amounts to 11,520 hours ÷ 960 mph = 12 months.

The Prophetic year is 11,520 ÷ 24 hours = 480 days

The sun revolves in its circuit around the universe or heavens according to what is stated in the scripture. The revolving sun is blown by its solar winds on a separate circuit from the moon respectively; and both at their own periodic times. The sun and moon was created to shine on the earth without the hindrance of celestial bodies or foreign objects blocking the rays and light that beams toward the earth, except when there is an eclipse. The above abstract is an accurate assessment and a realistic functioning of the earth on its axis in variance to what the scientific world has published in text book, journals and on the internet.

A degree is known as a plane angle at 1/480 degrees in measurement of a complete circle or rotation. The Prophetic calendar has 480 days in a year and 40 days in each month. The exponent is 480 degrees divided by 24 hours = 20 degrees along the longitude and latitude lines. An Accurate measurement can be taken directly from a location at the Dome of the Rock in Jerusalem at the center of the earth.

CIRCUMFERENCE OF THE EARTH:

The earth is divided up into 480 degrees. It is as follows:

From 0 degree east to south east = 60 degrees or 1,440 hours

60 degrees south east to south = 120 degrees or 2,880 hours

120 degrees south to south west = 180 degrees or 4,320 hours

180 degrees south west to west = 240 degrees or 5,760 hours

240 degrees west to North West = 300 degrees or 7,200 hours

300 degrees North West to North = 360 degrees or 8,640 hours

360 degrees north to north east = 420 degrees or 10, 080 hours

420 degrees north east to east = 480 degrees or 11,520 hours (starting to ending points)

SUMMER AND WINTER SEASONS:

Pragmatically, there are two seasons that the Creator of the universe has assigned to the earth - namely summer and winter. The summer season slowly becomes hot when the earth is starting to reposition itself at the beginning of the winter at 240 degrees. The earth is tilts to the left at an angle of 30 degrees. The sun makes its yearly revolution around the universe in the summer as close as 120 degrees south, changing the temperature to the hottest months in the year. Actually, summer begins in the first month, and it ends in the sixth month on the Prophetic calendar of 480 days. On the Gregorian calendar, the summer is recorded as starting in the month of mid-June and ending in September occasionally. The winter season begins in the seventh month, and ending in the twelfth month on the Prophetic calendar. On the Gregorian calendar the winter starts in mid- December at times, and ending on the 21st of March, depending on the year it is falling in.

In tropical countries there is summer, monsoon, or rainy seasons.

THE SUN'S RELATIVITY TO EARTH:

This is an accurate assessment at the time the sun sets, or the sun going down in the northern and southern hemispheres, east and west respectively. They are as follow:

Rises – 6:00 am

Overhead – 12 noon

Sets – 6:00 pm

In accordance with the earth's cylindrical shape, the above computations clearly indicate that all angles are equally divided. This approach is consistent with a detailed functioning of the sun and moon's gyration around the earth. However, there are dissensions within the scientific community as to the sun being fixed in the center of the universe, or a claim that the earth is rotating; and at the same time it is orbiting around the sun for 365 solar days. Actually the sun orbital cycle around the Cosmos is 480 days in a year.

THE WORLD'S GLOBE:

The present world's globe shows that the Antarctica (Antarctic Circle) is surrounded by the following waters:

- South Pacific Ocean
- Drake Passage
- South Atlantic Ocean
- Indian Ocean
- South Pacific Ocean

It's a possibility that the Arctic Ocean where Greenland is situated is linked with the Antarctica, and also teaming up with a steaming ocean current beneath the crust of planet earth; and it is known in Genesis as the fountain of the deep.

Genesis 8:2 stated that the waters beneath the earth, coupled with the rain that flooded the earth, lasted for 40 days and 40 nights. It clearly explains that when the flood waters were abated from above the earth's crust, it did return to its place of origin in the underworld. With an analytical reasoning, the melting of the Antarctica's is logically suggesting that the waters are surging along with the underworld ocean current - maybe it's the answer in determining what is causing the rising tides at the various locations. This had previously taken place as the case of Noah's flood in biblical times. Also, other things as new big ships or the dumping of waste matters, can be significantly contributing to the rising tides.

The above mentioned oceans and passage clearly separate the Antarctica from the rest of the civilized world. The questions are: is the Antarctica a separate entity or another place from our civilized world? Was the Antarctica ever fully explored, so as to determine any sort of life existence? Definitely, life does exist beneath the earth.

They are humankind or hybrids, but they are not associated with the human race. The question is: was the Antarctica topographically and fully measured in its longitude and latitude lines with a 360 degrees compass, to ascertain an accurate size of the earth?

UNCERTAINTIES:

If there is no substantial evidence to prove that the Antarctica was fully surveyed. How then that the latitude and longitude lines are unequal drawn on the map of the world? : Longitude N and S equaling 165 degrees; latitude E and W equaling 180 degrees; and a cutting edge that seemed to have occurred when the world map or the Globe was completed. The final map of the world is indicating that the Antarctica was partially mapped.

A NEW JERUSALEM:

Hypothetically, in the continent of Antarctica lies the unpolluted New Jerusalem covered with ice, even though scripture states that it will be descending from the heaven. Is it waiting to be submerged, air bound and taken to its new location after the melting of its ice? The present world atlas has nothing substantial on the cartographic areas of the entire continent of the Antarctica, which is located beyond the continent of Australia. Though the above diagram (360 degrees compass) has placed a partial longitudinal Antarctica, but this is only a tip of the iceberg.

Jerusalem is considered part of Africa in the center of the earth, and an ideal location for crafting a map of the world, to achieve an extreme accuracy and precision. In order to obtain meticulous details and avoiding mistakes is to use the 480 degrees compass, as the instrument in creating an atlas of the world. The 360 degrees compass well falls short of 120 degrees in making a complete circle. The current 360 degrees compass is disproportional compared to the 480 degrees compass that has the required amount of degrees on its chart.

The newly invented compass is able to accurately pinpoint the cardinal points, and it has the correct amount of degrees on its chart to compass a craft into any direction in the universe. In general, the earth is equally divided at 240 degrees each, 240 degrees in the Northern hemisphere, and 240 degrees in the southern hemisphere, totaling 480 degrees. This is in variance to a 360 degrees compass that gives measurements from the south latitude, North latitude, the equator and the South Pole, with each having a 90 degrees angle. The longitude is measured at 180 degrees to the east, and 165 degrees north from the Prime Meridian. Without a doubt the shape of the earth would show a bulge, looking like a sphere or oblate at its pole. The visible portion of the Antarctica on the current topographic map of the world shows that south latitude is 60 degrees, a clear understanding that the Antarctica was not measured to its utmost boundaries.

SYNCHRONIZED MATHEMATICAL FORMULA:

The scientific world regards the sun as being stationary in the universe, but the scripture states that the sun rises in the east, and sets in the west. As a matter of fact, a very powerful solar wind channels the sun in its designated circuit around the entire universe or world at a speed of 3,500,000 mph. It subsequently returns to a starting point in the east, via north bound. The sun routinely orbits the universe at 40 days each month (480 ÷ 12), in a year of twelve months totaling 480 days. The distance of the sun from the earth as it is gyrating around the universe is fluctuating, because it is revolving in an elliptical manner, and making it impossible to take an accurate measurement while in motion. The sun is journeying in its circuit, at 84,000,000 elliptical miles on a daily basis around the heavens and including the earth. The calculating time that the sun is providing sunlight to the earth is: 84,000,000 elliptical miles. It clearly demonstrates that some destructive forces are challenging or distorting God's creation, with Him constantly recreating the heavens and the earth.

A hypothesis is that the entire universes, including the sun which is travelling faster than everything else in the heaven is moving along counter clockwise with other planets, and it is traveling in an elliptical manner non- stop; and the earth is also part of that elliptical ring but it is moving counter-clockwise. The earth, as it is known is rotating on its axis at 960 mph. Any space craft that is penetrating into that circle in the universe which is swirling in an elliptical fashion will fall in line with the movement of the universe and causing a delusion that the sun is at a standstill position at all times. There is no single telescope that would be able to span the entire universe in determining its synchronized movements, its expansion, its direction of expansionism and the rate of speed it moving.

EARTH'S GEOCENTRIC:

The earth's magnetic core, measuring from the surface of the earth and from the Dome of the Rock is located east 480 degrees, minus 120 degrees depth = 360 degrees at magnetic north. Therefore, the density from the surface of the earth to its core is, 5,760 miles divided by half (2) = 2,880 miles. Whatever is in that core, it is generating heat in penetrating the crust of the earth. Job 28:5 states that as for the earth, out of it comes bread (food). Under it is turned up as it were fire.

EXPERIMENTING MOTION OF THE MAGNETIC FIELD

ABSTRACT:

From June 1st 2012, an experiment was conducted to monitor movements of the magnetic field.

EXHIBITS:

Bowl, water, needle made of steel (3), thin plastic float (3), 480 degrees chart, and a 360 degrees chart.

AIM

To demonstrate that the magnetic field is immovable beneath the earth's crust.

EXECUTION:

The needle placed within a thin plastic container in a bowl of water; it will immediately turns in the northerly direction.

ACCOMPLISHING RESULTS:

The results are revealed, that from the moment the experiment started, the needles immediately turn and they all were pointing in the direction of magnetic north at 360 degrees.

On June 21st, some sort of phenomenon took place, which caused the needle to move a couple of degrees counter clockwise up to August 31st, at 330 degrees, and decreasing by 30 degrees.

On August 1st 2012, the needles reversed its position to clockwise simultaneously, to 360 degrees north, reversing to its original position.

On Tuesday, August 7th, 2012 between noon and 430pm, the compass needles shifted to the north easterly direction at 390 degrees, an increase of 30 degrees on the 480 degrees compass chart.

On the 360 degrees chart, the needle moved in the direction of north east, a decrease of 30 degrees. On August 13th 2012, between midnight and 2: 00 am, the needles moved back to its original position at 360 degrees north.

On September 10th 2012, the needles slowly moved from 360 degrees north decreasing by 30 degrees September 21st 2012, the three needles in three different containers, moved back to magnetic north at 360 degrees.

It was observed that the sun rises in the east and goes down in the west in the same position on a daily basis. The shifting in degrees is taking place during the rotation of the earth on its axis. The earth tilts at a 30 degrees angle counterclockwise, which is to the left as indicated at the above mentioned times and date. What is being observed is that the sun at this time in the month of October 2012 is not overhead, but it is travelling to the right of planet earth, and measuring at 30 degrees away from the earth.

October 10th 2012 needles moved 30 degrees left of 360 degrees and again moved back to 360 degrees days after. On February 23, 2013, all three compass needles moved 30 degrees left of 360 degrees north to 320 degrees on the 480 degrees compass.

On March 3, 2013 the compass needles returned to 360 degrees north.

WATERS BENEATH THE EARTH:

Looking back into biblical times, the human race procured fresh water from artesian wells. Modern day man has explored new avenues, opportunities, techniques and innovative solutions, in order to obtain fresh drinking water for its population and other purposes. However, the earth was designed by the Creator, to receive a significant amount of rain fall to essential or designated areas. Subsequently, mankind has explored new avenues or opportunities, along with innovative solutions, to create involuntary rainfall to designated areas on land.

There are numerous complexities which may have occurred, along with inaccuracies navigating the high seas. Usually, unnecessary or excessive distances may have to be covered in getting from point to point, because of an unreliable 360 degrees compass, which does not have the required amount of degrees to accurately do the task. A 480 degrees compass is the appropriate instrument that may be able in achieving any sort of navigation, either by extending its capacity by going into another dimension in space, in the sea, on land and under the sea with a high degree of accuracy and precision. Scientific explanations given by the experts in the field of geology, geophysics or earth science may have deduced that the earth has a spherical shape and bulging at some point. There is uncertainty in determining the true center point of the earth, considering that the earth's entire land mass may have been an island, surrounded by approximately 75 percent of water, with only 25 percent land mass; a similar structure made up of the human body.

For over thousands of years, and within a twenty one year cycle of the moon, the earth maybe expanding from within its core. The land mass had expanded, volcanoes erupted, new island had submerged, landslides were on the increase, lands went under water, mountains crumbled, and the separation of land or continents took place. This is considered a progression and the growth of mother earth, the same as humans, humankind, animals, plants, and even rocks. However, all minerals shall be gone except the oil, which will remain for the purpose fueling the lake of fire as mentioned in the scripture.

ALTERNATIVES:

The premise is that people are very adamant or pessimistic about changing from their traditional way of thinking. They are very skeptical or uncomfortable with new discoveries made by the scientific community. At times, they would be willing to accept portent things from astronomers, astrologers or ancient people that were extinct or no longer unified on earth. In the scientific community or the academic world normally show resentment, when anyone is mentioning the name of God, they would strongly ridicule or degrade anyone who is suggesting that the universe was created by one spiritual God, and they would totally take Him out of the equation. One of God's attributes is forgiveness. He has an open door policy to anyone, in any language and to those who is sincerely seeking Him. This statement is not constructed with dogmatism, but rather intended to identify that man is being led astray by humankind from the time of creation, thousands of years ago. People having the spirit of God within them are being deceived by intellectuals; and some of the wealthiest and influential people on earth that has the spirit of Lucifer, that is of the world. Those that are in power have the master key to a double lock. When someone is caught in a snare, they are definitely trapped and cannot wiggle their way out.

The various Social Medias present their views in a logical and subtle manner to the masses that fundamentally accept, or trust what is being told to them. In some instances the story is not fully told or it is given with 60 to 70 percent inaccuracy, incomplete or totally inaccurate by some mainstream media.

EARTH'S LAYERS:

Many explanations or hypotheses were given in the eighteenth and nineteenth centuries by scientists, that beneath the earth's crust carries approximately three or more different layers of minerals, even though no human expedition had ventured deep down below the earth's crust.

In the twentieth century researchers made use of new technology or laboratory experiments in determining what's beneath the earth's crust. A scientific theory may have conclude, that their findings are based on seismic data, rock samples, or other laboratory testing's. However, the scripture describes that life, waters, fire and the various types of minerals are deep down in the earth as iron and copper, which some of the minerals shall be used in a destructive manner at the end of the age. One of the minerals is the oil that shall be fueling the lake of fire where all sinners will be thrown into to everlasting torment.

NAVIGATING THE UNDERWORLD:

A practical and workable solution in navigating the Underworld other than drilling from the earth's crust is to attempt the following:

- Construct a sub- craft made of 50% magnet and 50% other materials.
- Install infrared cameras, heat seeking instruments; and a nuclear power generated engine to navigate the craft back to base.
- Craft should be controlled by a robot, instead of a human being.

Sub- craft should be taken to a central point of the Arctic and Antarctica as a launching point of the mission and set to nose dive at 360 degrees north, in the direction of the magnetic field but, pointing towards 120 degrees, in the southerly direction by using a 480 degrees compass.

MAN'S FREE WILL:

God fashioned the cosmos including the earth for his purpose and for man. However, the mainstream scientists are very skeptical about spiritual things or a spiritual God, but they are having confidence in their own ability in discovering new thing in the universe. Man has a free will in making choices in his beliefs on how and when the solar system, including the earth was formed. Science estimated that universe started at an estimated 4.5 billion years ago, this is in accordance with their calendar and timeline.

Hypothesis: The underworld and its oceans can be merging with the oceans in the upper world. It is a possibility that the link maybe located somewhere in the Arctic Ocean or the Antarctica region.

The above presentation is not solely based on academic researching, but it's spiritually and an inspirational to an understanding as to how the earth is functioning from above and deep down beneath its crust.

360° MAP OF THE WORLD

The below map indicates that the center of the earth is somewhere in the Middle East. According to the 480 degrees compass, the center of the earth is Jerusalem located in Africa.

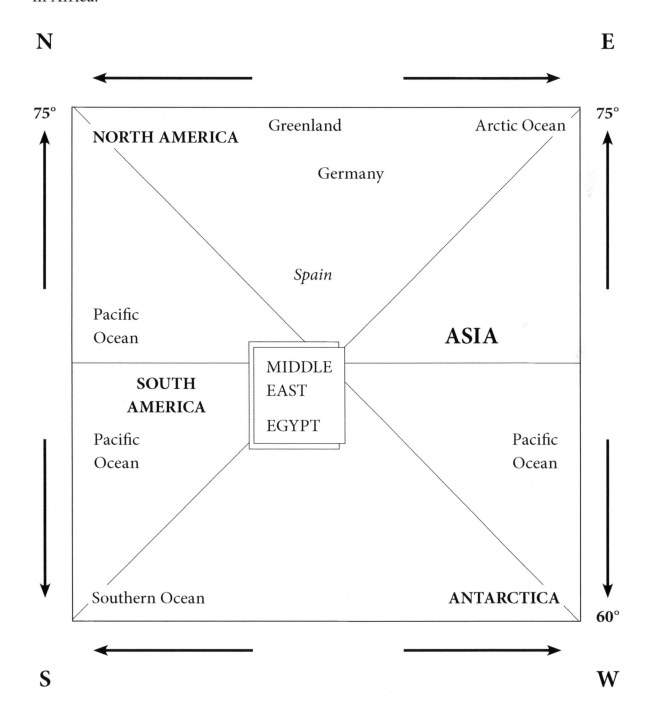

EARTH'S 480 DEGREES COMPASS:

Proverbs 8:27 states that God set a compass upon the face of the deep. This compass chart has a complete circle of 480 degrees. The earth also carries 480 degrees or 480 days in one year, and it works in conjunction with the two seasons – summer and winter; each carrying 240 days per season. Psalm 19:6 states that the sun starts its journey from the borderline of heaven, and nothing escapes its heat during the twenty four hours gyration around the galaxy.

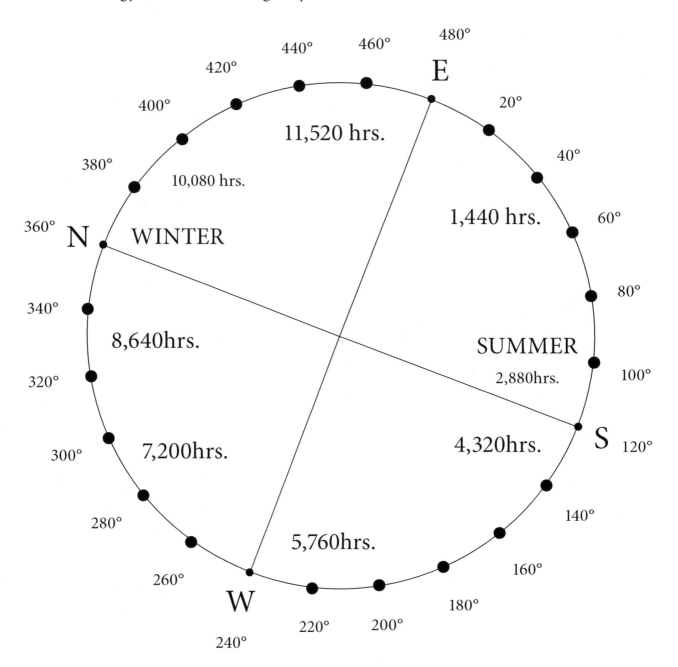

FIGURE B: 480° EARTH

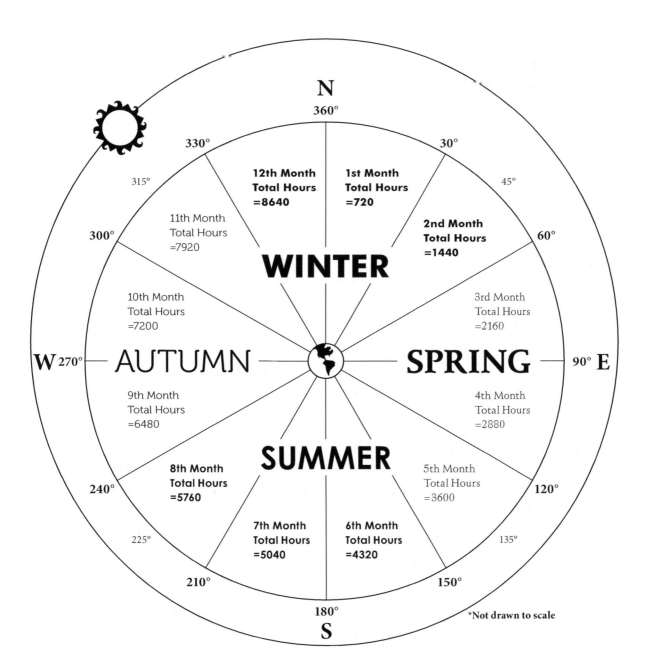

Figure A: An illustration of scientists complete circle of 360 degrees which, is short by 120 degrees or 2,880 hours to complete 480 degrees prophetic circle.

CONCLUSION

The biblical text, reinforces that there should be an adaptation of a Prophetic calendar. There are many examples in this book along with view-points, why the world needs to reform the Gregorian calendar. Incessantly, for over five centuries ago, the Gregorian calendar System has been the unofficial global standard recognized by the United Nations and other international bodies that are benefitting from the Gregorian calendar system. Overall the Universal Postal System, the Banking or Mortgage Systems, the IRS and the IMF are reaping huge profits from the present system, and it is also used for the basis of radio carbon dating. The calendar's system is not functioning as expected by its originators in the office of the Vatican. In this sense, an adaption of the Prophetic calendar will be beneficial to the world, and it may eventually initiate the following changes by sharing ideas, new technology, exploring new pathways and improving one's life.

Developed nations will be showing a genuine interest in the third world nations by offering free training programs in the field of agriculture; and promoting innovation instead of weaponry. It will be improving the educational system in many areas by encouraging individual growth and development. Therefore, in reforming the Gregorian calendar and replacing it with the Prophetic calendar, the world shall appreciate and they shall be surprised to know, that the people are not as old, but they are younger than their age.

Most text books will have to be corrected to fall in line with appropriate dates. Accurate documentations shall be maintained along with making the right prediction about future trends; or forecasting the weather with the avoidance of unnecessary mistakes and errors being made.

A very good role model is those astronauts aboard the space station that are representing their individual nations. Somehow they are promoting a strong working, supportive and collegiality in outer space. This professionalism engenders cohesiveness, and it is effectively and efficiently achieving measurable objectives, that are establishing set- goals for their individual countries. This contemporary impression is displaying a very good example of promoting or sharing ideas with each other on earth by instilling trust and unilateral cooperation.

Those astronauts on the space station have been recognized as good team players, and working similarly to a well-organized orchestra, that is playing musical symphony to the enjoyment, and in the ears of everyone.

The Prophetic calendar will be working in conjunction with the Jubilee calendar, but it is not required to be in-cooperated with any other calendar for either civil or religious purposes. It is not necessary to add or subtract a month, so as to prevent an adrift or variations from the season. It is extremely accurate, simple for the planner; and it will be effective in developing programs at a national level for an organizational requirement in achieving everlasting results. If the Prophetic calendar is adapted, it will not be necessary to turn the clock one hour forward and reverse it back an hour, so as to balance the daylight. With the introduction of the calendar, countries may well break even, in order to start a new year in the first month; and a month when all things were recreated on earth along with the universe.

The forty or more calendars are somewhat shorting approximately 115 to 120 days in a calendar year in comparison to the Prophetic calendar of 480 days. Their irregularity programs a society for fiscal and stressful problems on an annual basis, but it assists the Banking and the Mortgage Systems in generating quick turnover on interest rates, all to their benefits because of lesser days in the months.

The Gregorian calendar system's turn around time goes by very quickly in paying bills or debts, which is causing many people in becoming depressed. It is forcing businesses or an individual into more fiscal problems- more so breaking up a family; and following with other serious financial consequences. The Prophetic calendar will also act as a fiscal modification to mortgage benefits to both borrowers and lenders, that is, if it is properly regulated under the government's control. Perpetually, the calendar has no need to compete with other calendars, but it will stands on its own. It will be synchronizing with the seasons and the weather throughout the year. It has a positive role in simplifying one's life. It has the capacity to make an impact on many lives by contributing with its highly extreme accuracy and trends. It will not confuse the younger generation in the future. It will be overcoming obstacles that will eventually be achieving greater and everlasting results.

As this earth's aged-cycle is currently closing, we are invited to lay a new foundation from which to create a new order, with highly advanced technology and innovative thinking in this twenty first century. What are required from nations is to move forward with a new time, a new calendar and a new system, with the capacity of stimulating them in the sense of producing a new generation of innovative thinkers.

Having a new calendar and replacing the outdated and drifting Gregorian calendar is like uniting all cultures on the planet regardless of race or religion. If nations can only respect the calendar of the Creator, and adhere to His commandments or statutes, the lives of human on earth shall be less stressful, very simple, enjoyable and will cut loses. "Here all nations on earth, the Lord are our God the Lord is one."

A commitment to honor man's time and deadline on its nation's calendar is very stressful. It brings on the following stress: nervous tension; anxiety; annoyances; burnout; pressure; unproductiveness; mistakes; hypertension and other serious ailments, that could lead to disability and death. The reason why the Gregorian calendar is creating problems for a society is that, because the numbers of days in the month of September is shorting 10 days, November 10 days, June 10 and April 10 days each. In the month of February it is shorting 12 days; and the other seven months they are shorting 9 days each. Nothing on the Gregorian calendar is done in an orderly fashion, but it is confusing to many.

The Prophetic calendar highlights ways to be productive. It will be bringing about changes that can be dependable, accurate and effective to the planners. It will pinpoint overall assignments; and in doing so it shall create an accurate almanac in the years to follow. It shall be a very reliable instrument to the farmers in their sowing and reaping of crops. It is the most practical and simplifying system to the planner for any futuristic events or contingency planning for airports and highways snow removal. All annual events will consistently fall on the same date and time forever; same as with the birth date and the death of all humans.

Man was created in the mode of freedom, peace and holiness, but on the account of Adam and Eve who had sinned against God, brought about pain and sufferings unto the future generation. All men have incurred a loss as a result of man falling from grace. Man has a choice within himself to do good or evil, and ultimately he has to make choices throughout his life. Life on planet earth is fairly good, yet painful, until our journey to death. Humans have to acknowledge that God's word is a proclamation with having the greatest power; and it is very influential against anything. Whenever He executes or gives a command to anything, it obeys and it is done. Psalm 2:10. With the introduction of a new World's calendar, the United States of America maybe continuing in achieving its leadership stature globally, but not based on economic, cultural, democratic or military superiority, but by influencing other nations in accepting a new World calendar. The United States of America's inspirational visions, and the sharing of new technology with the stabled or undeveloped nations, may promote a strong bond with those countries. Making resourceful decisions, visualizing a short and long term consequences may elicit confidence or emanate a catalyst of trust with the utmost success and respect from others; and not expecting any negative setbacks from the other Nations.

On the flip side, nuclear weapons do not radiate authority or display strength, but they are causing distrust among many nations, which is ultimately driving them into developing or upgrading their nuclear weapon capabilities. If some of the other countries in South America, Saudi Arabia, Egypt or countries in the African continent acquire the technology and the capability in developing nuclear weapons; it would be driving fear in the entire world. They may be feeling very insecure in their future; or the future of their children and children's children.

Religious bickering, invidious behavior, violence, hate or dominance is not avenues to a new order of the world. It is a double –edge sword against the nature of the Almighty God. Nations should be attempting to make an effort in preserving a workable relationship with each other, to maintain peace among them. The world may be better off if people have tolerance, love and respect for each other; and in displaying ingenuity along with some initiative in resolving problems. It will benefit every nation and everyone to live at a higher level of consciousness and awareness of the violence that is plaguing the nations in the world; and to do something about it by curbing those tyrants who are responsible.

The scripture clearly implies that everyone should be aware of false prophets. Those teachers, preachers, parsons and others that are teaching their followers about violence and hate are conceived as segregates, being greedy wolves in sheep's clothing and a false prophets or false Christ. And some religious leaders with a flattering title are not sincerely representing the true word of God. But they are displaying a dynamic or charismatic leadership performance to their followers. Matthew 7:15; Mark 13:22; Revelations 13:13.

The most effective way in accomplishing peace is, that all religions and governments honor their creator on one accord at the duly appointed times, to ascertain a specific or positive results and spirituality. The leaders and the people on planet earth should be making some time in meditating unto the Lord. The world that we live in is not operationally functioning by itself, but it is totally in control by the creator who designed it. Righteousness and spirituality are for those that sincerely need it when seeking the Almighty God that created the worlds. Religions are for those who are desperately seeking to be rewarded for their works in their lifetime. Religious ideology or social order may emanate within the three main religions by working diligently, harmoniously and constructively to avert conflicts. No individual nation should control and decides as to when, where, how and why to use nuclear weapons on another nation. None of the nations should be working in concert in planning to nuke another nation, because of their indifference or any inappropriate languages used against them; or distasteful words that may be upsetting to their religion. The scripture mentions that a few countries will be having an illegitimate government and leaders that shall be coming up against God's people in the end times. It's implying that those extremists, terrorists, tyrants or radicals shall be partaking in the final battle between good and evil.

Straightforward questions in reference to the Gregorian calendar are - why take an engine built in the year 1582 then install it in a 2013 new model car and place it on the market for sale? It's the same as the Gregorian calendar being adapted in the year 1582, and the world continues to use it in the twenty first century in a new developing world. Smart are the governments and politicians of nations, but wiser is the population, that pinpoints the special interest groups and selfish individuals that lobby for positions, but not to the best interest of their people.

The initiation of a new Global calendar will enable the governmental bodies to observe and recapture all opportunities of illegal activities at an early stage. Creating new programs, and closely monitoring the terrorists, all its users or surfers that are navigating the World Wide Web - it will bring about stability to the nations. A united body of trusted and technologically developed nations should be controlling and monitoring the World Wide Websites. This system may be able to prevent terrorists from initiating and finalizing any plots in destroying peaceful and democratic nations. Overall the above recommendations given are only a few practical objectives to a sound strategic plan. It will apply to those nations that are unequipped with sophisticated and new technology, in preventing any terrorist unexpectedly breaching their nation's security system. It is a tool, either preventing any internal or external aggression from cyber terrorism that are premeditating in hurting its nation or other nations.

The final power of the governmental body in the new world to come shall be, that anointed one of a bloodline from king David sitting on His throne in the new Jerusalem; and governing a newly created earth. All potential changes that are required to be made on the Prophetic calendar are a prelude in anticipating the reign of the anointed one from the dynasty of King David.

In deprogramming the human mind to see the true light of the world, and the future of humanity, is to adapt a simple and usable calendar. At some point nations would have to walk past the burial sites of the Gregorian and Lunar calendars, because they are becoming obsolete, and the makers are exhausting their means of doing adjustment with their calendar.

The momentums of changes in these nations are in the spirits of Shem's descendants, with most of them being Jews, some are Christians; and so are many Arabs in the world. A continuing disputing of land and other religious conflicts are anticipated to be escalating in the in the continent of Africa. There is a formula for the end of the age, where it can be identified on the Jubilee calendar. Tribulations or the various natural disasters shall accelerate, while innovative technologies shall be augmented continuously or either to attain greater contribution to the world or for self- destruction. However, decisive planning and finding solutions to feed the fast growing population in the world, should be a prioritizing objective for all nations to speedily produce their own food in abundance.

The God that created the universe is fully in control of the earth, beneath the earth with its humankind, which is monitoring man's activities at all times. He shall create new charismatic and flamboyant leaders of nations, to accomplish changes, and evoke a new revival in the world, in accordance to His designed plans for the human race and mankind. All things which are done with no good intentions shall totally collapse. Young men will have dreams about things to come, and their dreams shall truly come to pass. Only those who have the spirit of God within them shall have positive dreams, unless God has an important message for an evil- doer, then He would give it in a dream or a vision, either once or twice to that individual.

Competent leadership with future vision of development and economic growth should be using the Prophetic calendar as a jump start to a new order, and to improve the various operations as a solution to resolve the agricultural problems.

The Prophetic calendar's main purpose is to give the proper and accurate time line of one's age, and the age of everything else in the universe including planet earth. Disrespecting or super imposing religion on each other and not putting aside petty indifferences will not effectively develop cohesive solutions to solve problems to benefit the world. Man- made religion and their clear vision of set- goals are to dominate or display power by controlling the masses - even in attempting to dominate the world.

Some motivational speakers in authority are relaying messages from the Creator to their congregation, or the Medias by relaying messages out of context, and giving their own opinions about things in the scripture. At times they take out a scripture verse from the bible, using it to their own advantage or benefit in getting a message over to their congregation. The leaders of religions are limiting their congregation on a need to know basis. They may be fearful in telling the entire truth, because of a reprisal from the opposition or evil- doers. No one has all the answers or symptoms to the chaotic situation in the world, but the one and only God of the universe.

God's chosen people who were given a divine name shall be part of the tribulations or changes, that is, if they are still alive during the tribulation periods. However, humankind will be witnessing and experiencing all the sufferings on earth until the end of the age. All things as the sun, the stars or the hosts of the heavens and the space station shall either be brought down to earth, or it will be disabled. Those things below the seas as the sea monsters, leviathan and the underwater vessels shall be submerging. Hell from beneath the earth will be surfacing, and it shall be merging with the lake of fire at the coming of the Messiah; and concurrently with the resurrection of the dead. **Isaiah 14:9, 15: Matthew 11:23**.

Peace on earth shall be achieved as the entire world will be witnessing, or shall be enduring the natural disasters, pestilences and plagues that will be destined for mankind. All flesh shall be destroyed in the end, but the chosen ones shall be transformed from the flesh to angelic beings in the twinkling of eyes, and without any pain or sufferings. Those intense tribulations which are mentioned shall be ongoing in the latter days, but not any time in the near future as highlighted on the Jubilee calendar.

Tribulation will continue through year – to – year in natural disasters. In Revelation 2:10, it gave a warning that the devil shall cast some of the righteous people into prison. It is a possibility that one may not be having a trial in the courts. The people who do not accept the mark of the beast will be thrown into the jail without a trial; and that is the ten days tribulation and tortures mentioned in the scripture. People who are not learning from their past mistakes- history tends to repeat them.

In order to determine an accurate start-date of the Prophetic calendar; its transference or conversion would decisively take place in the first month on the Prophetic calendar. On the Gregorian calendar, the transformation process would be either in the month of March or April, depending on the adaptation year of the Prophetic calendar.

Prophetic Timeline:
 Birth of *Yehshua* – The 13th of the first month in the year 3949
 Death of *Yehshua* – The 13th of the first month in the year 3979
 (A biblical day begins at sundown)
 Age of death – 30 years old
 Destruction of second temple – 4009 years
 From destruction of Temple to date – 1,500 Prophetic years
 From Gregorian years to Prophetic years – $2013 \times 365 = 734,745$ days
 Conversion – $734,745 \div 480$ Prophetic year = 1,530.7 year
 $1530 + 3979 = 5510$, (a Prophetic year, or 2013)

Gregorian Timeline:
 Death of *Yehshua* – 1 AD (3979)
 Prophetic plus Gregorian = $3979 + 2013 = 5992$ Gregorian years indicated on Jubilee calendar

End Times:

> *Yehshua* to the end times – 6999 – 3979 = 3,020 years
> Present time to end times – 6999 – 5510 = 1,489 years
> From *Yehshua* death to date – 1,530 years

The moon matrix shows that a twenty one year prophetic cycle is about to conclude between December 2013 thru the month of March, in the year 2014. This will be after the moon and sun have completed their 340 times orbit- cycle around the earth. Because of the complexities of the Gregorian calendar, this date may not be accurate. At the end of a twenty one years cycle, the sun and the moon that were created on the said day, shall work simultaneously in completing a reversal- cycle in the Prophetic year of 5509 (2013). Luke 8:17 states that nothing is made secret, that is not made manifest; or neither anything hidden that shall not be known and come abroad.

Before any hard and fast conclusions are drawn, be aware that this book is not intended to impose anything on religions, the ruffling of feathers of professionals, the preachers or by using any disparaging or degrading statements, finger pointing at the scientific community, the IRS, the IMF, or in attempting to convert anyone of their beliefs. Its purpose is merely to bring about the awareness's that God is fully in control of His creation, and His entire universe along with its hosts. All living soul shall be judged by Him with His twenty four elders, and along with the saints at the end of the age.

REFERENCES:

The following scripture verse references have given detailed and a blue print specifications on how the heavens and the earth were created, the motions of the sun, moon and the earth and who created everything that existed and existing. These references are a validation to the title of the book- YOU ARE YOUNGER THAN YOUR AGE.

1John 3:9. Who so ever is born of God does not commit sin; for his seed remains in him.

Job 26:7 states, that God stretched out the north over the empty place, and hung the earth upon nothing. Psalm 89:11.

Job 26:13 states, that by the spirit of Yahweh, he had formed the heavens, and that his hand formed the crooked serpent [Lucifer].

Job 22:12, 14 states, that God is somewhere high up in the heaven above the stars, and that he walks in the circuit of heaven.

Enoch 72:37 states, that the sun rises and sets; it neither decreases nor rest, but runs day and night in its chariot. Its light is seven times brighter than the moon, but in size the two are equal.

Joel 2:10 The earth shall quake before them. The heavens shall tremble; the sun and the moon shall be in darkness, and the stars shall withdraw their shining. Isaiah 13:10.

Matthew 6:45 states, that God makes the sun rise for evil and for good.

Hebrew 11:3 states, that through faith, the worlds were framed by the word of Yahweh, so that things that are seen were not made of things which do appear. Hebrew 1:1.

Proverbs 8:27 states, that when God had prepared the heavens.............

2 Peter 3:10 states that the heavens shall pass away with a loud- noise and the elements shall melt with fervent heat. Also, the earth and everything within it shall be burnt up.

2 Peter 3:7 states, that the heavens and the earth, which are now by the same word are kept in store and reserved unto fire against the Day of Judgment and perdition of ungodly men.

Psalm 19:4-6 states that the sun goes forward from heaven, and its circuit to the end and nothing escapes its heat.

Malachi 1:11 states that from the rising of the sun, even the going down of the same, God's name shall be among the Gentiles.

Joel 2:30 states that God will show wonders in the heavens and on the earth with: blood; fire and pillar of smoke.

Deuteronomy 10:14 states, that the heaven and the heavens are the Lord's; also the earth with everything.

Revelations 12:12 states that, those in the heavens which dwell in them should rejoice.

Ecclesiastes 1:5 states that the sun also arises; the sun goes down and hastens to its place where it arose. Psalm 19:4.

Isaiah 48:13 states that God's right hand had laid the foundation of the earth and his right hand has spanned the heavens. Psalm 102:25; Nehemiah 9:6.

Isaiah 43:10 states that you are my witnesses and my servant who I have chosen that you may know and believe in me and understand that I am he. Before me, there was no God formed, neither shall there one after me. Isaiah 44:8; 41:4.

Psalm 68:4 states that God rides upon the heavens.

Psalm 97:2 states that clouds and darkness are around him.

Psalm 104:5 states that the foundation of the earth shall not be removed ever.

Psalm 144:5 states that Yahweh should bow his heavens and come down. Isaiah 45:8.

Isaiah 50:3 states that Yahweh clothed the heavens with blackness.

Isaiah 14:13 states, that Satan said within his heart, that he shall ascend into heaven to exalt his throne above the stars of God, and he shall sit upon the congregation in the sides of the north.

PSALM 115:16 states that the heaven, even the heavens are the Lord's, but the earth has he given to the children of men

Psalm 148:4 said to praise him you heavens of heavens and the waters that are above the heavens.

Psalm 33:6 states that by the word of God were the heavens made and all the hosts of them.

Psalm 104:3 states that he covers himself with light as with a garment and stretches out the heavens like a curtain.[The heavens continue to expand].

Isaiah 66:1 states that the heaven is God's throne and the earth is his footstool.

Isaiah 66:22 states that as for the new heavens and the new earth which he shall make shall remain before him.

Psalm 68:33 states that to him that rides upon the heavens of heavens which were of old.

Isaiah 60:20 states that at the end of the age, the sun shall not go down or the moon withdrawn, because he shall be the everlasting light.

Jeremiah 10:11 states that Yahweh made the heavens and the earth and they also shall perish from the earth and from under the heavens.

Revelations 12:3 states that no man in heaven or on earth, neither under the earth was able to open the book. [Our neighbors are right under us- no need to travel into outer space].

Isaiah 55:9 states that for as the heavens are higher than the earth, so are God's ways.

Psalm 8:3 states that God considers the heavens, the work of his fingers, the moon and the stars that he has ordained.

LIST OF NATURAL DISASTERS – WIKIPEDIA, THE FREE ENCYCLOPEDIA:

1. History of science, and technology in China – Wikipedia, the free encyclopedia
2. Gregorian calendar – Wikipedia, the free encyclopedia (http/en.wikipedia.org/wiki/Gregorian calendar)
3. Weather history for the JFK NY.USA
4. Jude 6 states that the angels had left their estate that He had reserved in everlasting chains under darkness unto judgment day.
5. Revelation 18: 4 says that a voice from the heaven saying, come out of her my people that you may not be partakers of her sins that you may not receive her plagues.

6. John 8:39; Romans 9:6; Galatians 6:16 says that they are not all Israel which are in Israel.

7. Corinthians 15:38 states, that God gives a body as it has pleased him.

8. Psalm 104:30 says that God sends forth your spirit; they are created and you replenish the face of the earth.

9. Psalm 147: 4. God can tell the number of stars and he calls them by name. Isaiah 40:26.

10. Psalm 147:16, 17. God gives snow like wool. He scatters the hoarfrost like ashes. 17. He cast forth ice like morsels. Who can stand before his cold?

11. Psalm 148:2- 12. Everything in this universe praises God in their own way; even the sun, moon, the waters, the snow, the animals, the dragons, creeping things, the birds, the trees, the stars, fire, hail vapors storms mountains hills, kings, princesses, young and old.

12. Proverbs 6:16-19. Six things that the Lord hates, the seventh is abomination unto Him. A proud look, lying tongue, a murderer, a wicked person, a mischievous person, a person bearing false witness.

13. 1 John 3:9. Whosoever is born of God does not commit sin; for his seed remains in him and he cannot sin, because he is born of God.

14. John 10:34. You are Gods, all of you children of the most high. But you shall die like men, and fall like one of the prince. Psalm 82:6, 7.

15. Revelation 3; 9. I will make them of the *synagogue of Satan*, which say they are Israelites and they are not, but they do lie. Behold, I will make them come to worship at your feet; and then they should know that I love you.

16. Romans 1: 23, 25. They are serving creatures more than the Creator.

17. Matthew 5: 17. Think not that I come to destroy the law or the Prophets; I come to fulfill.

18. Isaiah 55: 8, 9. My thoughts are not your thoughts, neither are your ways, my ways, said the Lord. For as the heavens are higher than the earth, so are my ways.

19. Isaiah 40:4; 45:2. Every valley shall be exalted. Every mountain and hills shall be made low and rough places a plain.

20. Proverbs 15: 1. A soft answer turns away wrath, but grievous words stir up anger.

21. It is an abomination unto the Lord for women going with women instead of men; and men going with men and lying in bed instead with women.

22. 1 Corinthian 15: 40. There are terrestrial bodies and celestial bodies in the universe and they both magnify the Almighty God.

23. Psalm 139:1-4; 7-10. The Lord knows and sees all things -nothing is escapable from Him.

24. Psalm 82:6. The children of the Lord are all Gods, but they will die like men.

25. Genesis 1:31. Everything on planet earth was made very well. He made all riches and minerals of the earth. P104:24; Psalm 40:5.

26. Isaiah 30:26. The light of the moon shall be the light of the sun (seven times).

27. Isaiah 60: 19. The sun shall not be anymore the light of the day or the moon at night. The Lord will be the everlasting light. Revelation 21:23.

28. Isaiah 28:2. The Lord has many paroxysms, storms, violent commotions of hail, mighty tidal waves of water overflowing the lands, and things He has in reserve to cast down to the earth.

29. Isaiah 43:7. Everyone that is called by His name, He has created for His resplendence; for His prosperity and for His achievement.

30. Psalm 40:2,3; Revelation 5:9. Only His chosen ones shall know the new song.

31. Isaiah 51:19. These things are predestined for planet earth: sadness; destruction; famine and killings.

32. Isaiah 13:13. Only the Lord can shake the earth so that it can be removed out of place.

33. Amos 8:9. At the end times, the Lord shall cause the sun to go down at noon and He will darken the earth during a clear day.

34. Luke 23:44; Mark 15:33. Yehshua died on the cross in the darkest hour at 9:00 pm. So then, His return shall be at 9:00 pm during the hours of Passover on the Prophetic calendar and not the Gregorian calendar.

35. Jeremiah 8:3. At the end of times men shall chose to die, rather than to be bearing the torture of the tribulation. Hosea 10:8; Luke 23:30.

36. Revelation 20:7. When Satan's 1000 years has expired in the year 6992 he shall be set free for a short time.

37. Revelation 20:4. Those who do not receive the mark of the beast on their forehead or hands shall live with Yehshua for 1000 years.

38. The angels had left their dwelling in heaven.

39. The seed of David shall be triumphant in opening up the sealed book to let loose the seven plagues.

40. 1 Corinthians 10: Those in the wilderness drank of that spiritual Rock that followed them for 40 years, and **that Rock was *Yehshua.***

41. 1 Corinthians 10: 21. Do not participate at the table of the Devil and also participate at God's table.

42. 1 Corinthians 14:34, 35. Women cannot be preachers in the church, because it is not permitted.

43. 2 Corinthians 5:1. **The new Temple in Jerusalem shall not be built by hand**.

44. Luke 7:28. John the Baptist was the greatest prophet ever.

45. Luke 11:18. Satan is not divided against himself. So then one should not use evil to fight evil, because it will invite more evil.

46. Luke 12:49. Yehshua did not attempt to bring about peace on earth, but to cause division. This can be recognized among the divisions of races, religions and ethnicity.

47. Daniel 12:1. Intensified tribulation will be upon the earth as was never seen before by man, but the saints or chosen ones written in the book of life shall be set free.

48. Ephesians 4:9. Yehshua had descended into the underworld to set free the saints. They were subsequently taken to paradise after Yehshua had ascended to heaven. Luke 23:43

49. Genesis 28:22; 31:13; 35:7. The corner-stone of the world is located in Jerusalem. God's future temple shall be resting on it. The place was once called El-Bethel a place where Jacob had rested after fleeing from his brother Esau. Also, a place where Abraham was blessed by God after defeating the five kings. Genesis 14:18-20.

50. Genesis 33:18. Jacob had journeyed to Shalem, a city of Shechem in the land of Canaan. Jacob later erected an altar there in Shalem and he called it El-loheisrael.

51. Titus 1:13, 14. Yehshua said that wherefore rebuke them accurately and be exact to the point that they maybe sound in faith. They must not pay any attention to Jewish fables and their commandment of men that turned from the true gospel.

52. Matthew 6; 5-7. When you are praying you shall not be as the hypocrites are, for they love to pray standing in the synagogues and in the corners of the streets, that they may be seen by men.

53. When you pray, use not vain repetition as the heathen does. Matthew 6:7

54. Noman can serve two masters. Matthew 6:24.

55. Matthew 12:26. If Satan cast out Satan, he is divided against himself. So then, how shall his kingdom stand?

56. Who so ever speaks a word against the son of man, it shall be forgiven, **but who so ever speaks against the Holy Ghost, it shall not be forgiven**, neither in this world or the world to come. Matthew 12:32.

57. Matthew 12:36. Every idle word that men shall speak, they shall give an account thereof in the Day of Judgment.

58. When the unclean spirit is gone out of a man, he walks through dry places seeking rest and finds none. Then he goes and takes with himself seven other spirits more wicked than himself and they enter and dwell there. The state of the man becomes worse than the first. Even so shall it be unto the wicked generation. Matthew 12:43, 45; Mark 5:9; Luke 11:24.

59. Matthew 22:37-39. The first greatest commandment is you shall love the Lord with all your heart, all your soul and with all your thoughts. The second commandment is you shall love your neighbor as yourself. Leviticus 19:18; Mark 12:31; Galatians 5:14.

60. Matthew 17; 12; 11:13-14. **Elijah** was reincarnated as John the **Baptist**.

61. Isaiah 13:9 Hell from beneath shall move to meet you. It will stir up the dead.

62. Isaiah 34:4. All the hosts of heaven shall be dissolved and the heaven shall be rolled together as a scroll. All their hosts shall fall down as the leaf off from the vine.

63. Isaiah 13:13. I will shake the heavens and the earth shall remove out of place.

64. 1 Corinthians 2:12. We have received not the spirit of the world, but the spirit which is of God.

65. Zechariah 5:2-4 Curse travels all over earth, enters the house of the thief and those swearing falsely.

66. Revelation 5: 3. No man in heaven, neither on earth, neither **under the earth**.

67. Isaiah 30; 26. The light of the sun shall be seven times as the light of seven days.

68. Mark13:22. False Christ and false prophets shall rise and shall show signs and wonders to seduce, if it were possible even to the elect.

69. Job 38:7. Where upon the morning stars sang together.

70. Job 38:38. When the dust grows into hardness, and the hardness clumps together.

71. Job 38:24. By what way light separates which sends the east wind upon earth?

72. Luke 15:10. There is joy in the presence of the angels of God over one sinner that repents.

73. Luke 16:15. God knows your heart.

74. Isaiah 60:19, 20. The sun shall be no more the light by day. The Lord shall be the everlasting light. The sun shall no more go down neither shall your moon withdrawn itself.

75. Psalm 148:1-14. All things that God is creating praise Him.